IMAGES
of Wales

RHONDDA COLLIERIES

The emerald greenness of the meadows in the valley below was most refreshing. The air was aromatic with the wild flowers and mountain plants. Charles Cliffes description of the Rhondda Valley before the arrival of King Coal.

Acknowledgements

I would like to thank the following people and organisations for their help and enthusiasm during the research and compilation of this book:
Mr T. Lundregan and Mr H. Shadbolt, Mr Philip 'Cuz' Madigan, Trealaw; Mr G. Gregory, Clydach Vale; Ms Kay Warren Morgan, Mrs Stephanie Thomas and Mr Nick Kelland of Treorchy Reference Library; Mr Cecil Thomas, Gabalfa, Cardiff. Ms Lynfer Roberts, Tourist Development Officer for Rhondda, whose help and guidance during the research period is greatly appreciated.

I also wish to thank the Staff at the Glamorgan Archives, Cardiff, the National Museum of Wales, Cardiff, and the Coal Authority Mining Records Office, Burton on Trent.

Compiled by
David J. Carpenter

First published 2000

Tempus Publishing Limited
The Mill, Brimscombe Port,
Stroud, Gloucestershire, GL5 2QG

ISBN 0 7524 1730 4

Typesetting and origination by
Tempus Publishing Limited
Printed in Great Britain by
Midway Clark Printing, Wiltshire

In memory of my father,
David William Carpenter,
who died through the miners' dust
and to those miners who suffered
and are still suffering.

Contents

Bibliography

E.D. Lewis, *The Rhondda Valleys*
Roger Williams & David Jones, *The Cruel Inheritance*
John Cornwell, *Rhondda Collieries, Vol.1*
Ken Llewellyn, *Disaster at Tynewydd*
Michael Ayers, *Masters of the Coalfield*
Ken Hopkins, *Rhondda Past & Future*
Dr. W. Gerwyn Thomas, *Welsh Coal Mines*
Elizabeth Phillips BA, *A History of the Pioneers of the Welsh Coalfields*
Ray Lawrence, *The South Wales Coalfield Directory*
Bill Jones & Beth Thomas, *Historical Glimpses of life in the Welsh Coalfields*
Dewi Davies, *Welsh Place-names and their Meanings*
Richard Watson, *Rhondda Coal, Cardiff Gold*
Hywel Francis & David Smith, *The Fed*
R. Page Arnot, *South Wales Miners 1898-1914*
R. Page Arnot, *South Wales Miners 1914-1926*

Newspapers
Aberdare Times
Cardiff Times
Cardiff & Merthyr Guardian
Glamorgan Times
Merthyr Express
Porth Gazette
Rhondda Leader
South Wales Echo
South Wales News
Western Mail

H.N.S.O.
Reports on Colliery Explosions:
Dinas 1844
Cymmer 1856
Ferndale 1867
Ferndale 1869
Pentre 1871
Tynewydd 1877
Dinas 1879
Naval 1880
Naval 1884
Mardy 1885
Wattstown 1887
Tylorstown 1896
Clydach Vale 1905
Wattstown 1905
Glamorgan (Scotch) 1932
Lewis Merthyr 1956
Clydach Vale 1965

Introduction

With the discovery of King Coal, the exploitation of coal in the early part of the nineteenth century and the subsequent industrial developments of the South Wales valleys began.

In 1911 the population of the Rhondda Valleys was approximately 150,000. Of those 42,000 were men and boys employed in fifty-three collieries, providing approximately 1/6 of all the coal produced in the South Wales coalfield.

In recent years, the decline of the Welsh coalfields in the 1950s and the inevitable closure of the valley pits resulted in mass unemployment as there was virtually no alternative work unless one was prepared to travel long distances, or move out of the area altogether. Hundreds of miners were either put on the dole or, because of their age, pensioned off.

Efforts were then made to bring alternative employment to the valleys. The collieries were dismantled and the land either just landscaped or developed to set up factory units. The main aim of the development, however, was to improve tourism and leisure.

Today, however, with the loss of these collieries, we may want to know more of their history, their tragedies, the developments of their sites.

In order to preserve memories of such a relevant part of our history, I decided to research this book so that people will be aware of the past and perhaps will appreciate the future.

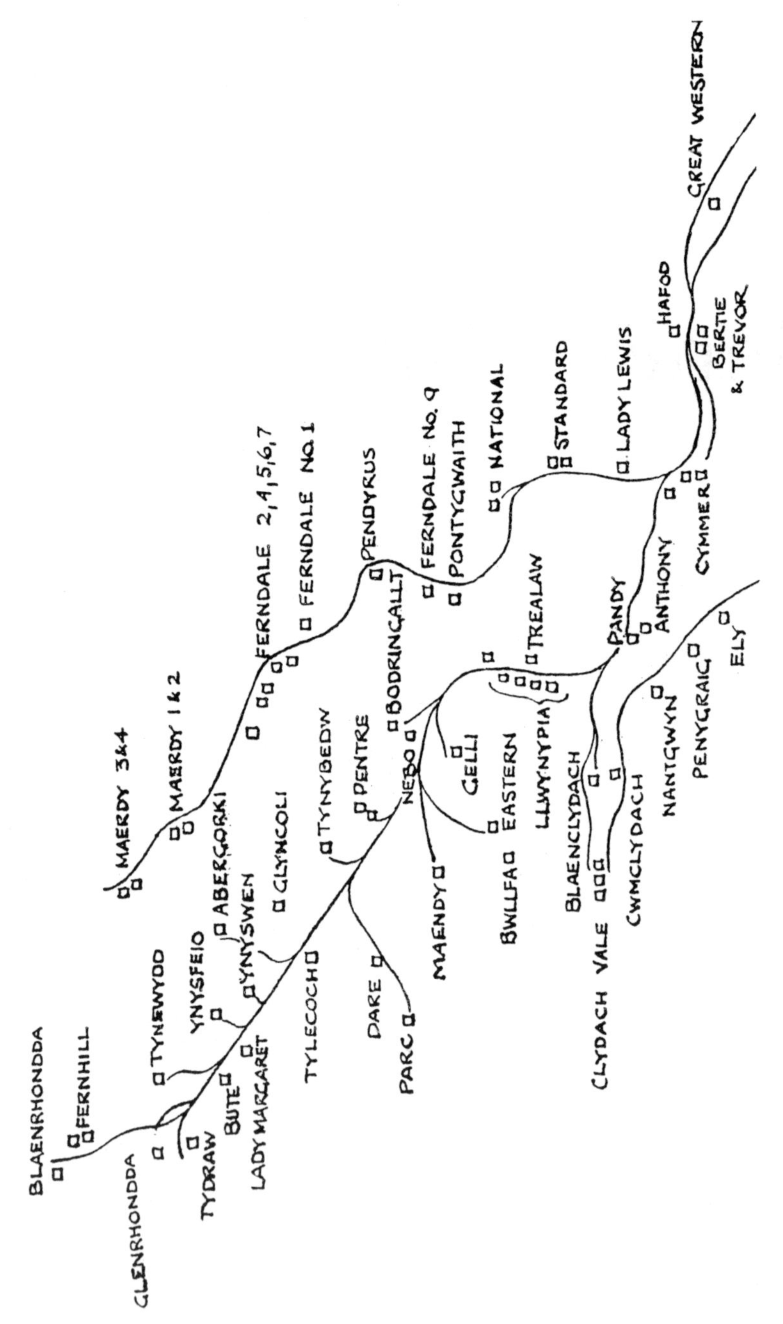

A map of Rhondda Collieries from 1913.

One
General Interest

Glossary of Terms used in the Mining Industry

District: Area of a coal seam isolated by means of barriers of untouched coal to ensure safe working conditions.
Heading: A main roadway in or out of a district.
Timbering: The fixing of timbers to keep the workings free from falling debris.
Cuttings: Removal of coal from the main seam.
Ripping: Cutting under and into the main seam to remove large quantities of coal.
Gobbing: The packing of cavities with waste after the removal of coal.
Dead/Day Work: Other duties such as labouring, cleaning out, preparing timbers.

The Rhondda Golf Club

The Rhondda Golf Club was formed in 1910 and later an eighteen-hole course was established on land at Penrhys. Popularly known as Penrhys Golf Club, the club perched 1,000ft above sea level and was formed from a typical cross-section of public life in the valley at that time, with miners mingling freely with the local professional and business people, all sharing an interest in the sport and the welfare of the club.

As a tribute to the closeness of the mining communities of the Rhondda, in 1910 the Rhondda Golf Club named each hole of the course after a local Colliery. The colliery could either be seen from the Tee-Off, or the driveway was in the direction of the colliery site.

The Colliery name attributed to each hole is as follows:

1. Tydraw
2. Parc & Dare
3. Pentre
4. Gelli
5. Glamorgan
6. Cambrian
7. Naval
8. Dinas
9. Cymmer
10. Lewis Merthyr
11. Bodringallt
12. Standard
13. Cynllwyndu
14. National
15. Pendyrus
16. Pontygwaith
17. Tylorstown
18. Maerdy

Due to further developments in the coalfield, some of these collieries were renamed: for example Pendyrus became Ferndale No.6, Pontygwaith became Ferndale No.8, Cynllwyndu first became known as the Pontygwaith and later Ferndale No.8.

William Abraham (Mabon) MP (1842-1922). A native of Cwmavan, he was appointed Miners' Agent to the Cambrian Miners' Association in 1877. He also represented the Rhondda as their first Member of Parliament from 1886 to 1920.

Typical colliers' checks used in Rhondda Collieries. When starting their employment at the local colliery, the miners were issued with a safety lamp and a lamp check, both with the same number. While the check was personally kept by the miner, his lamp was kept in the Lamp Room at the colliery. At the start of his shift, he would replace his lamp with the check, thus indicating to the time-keeper that he was at the pit. At the end of his shift he would replace his lamp and remove his check indicating this time that he had left the pit and returned to the surface. In the event of an explosion or any other tragic circumstances, it would indicate those who had become victims of an accident.
(Photograph by courtesy of Mr T. Lundregan, Trealaw)

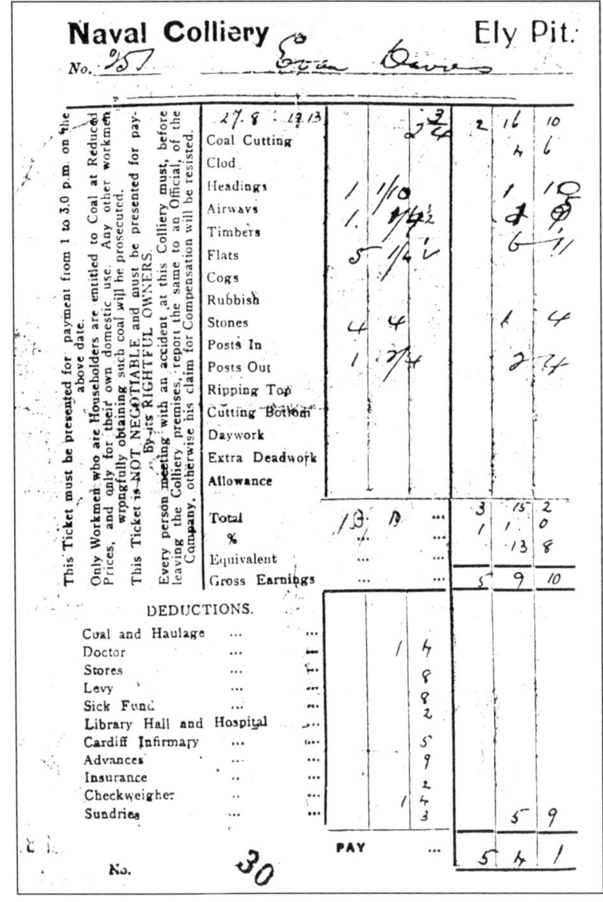

Naval Colliery Ely Pit.

No. 951 Evan Davies

This Ticket must be presented for payment from 1 to 3.0 p.m. on the above date.
Only Workmen who are Householders are entitled to Coal at Reduced Prices, and only for their own domestic use. Any other workmen wrongfully obtaining such coal will be prosecuted.
This Ticket is NOT NEGOTIABLE and must be presented for payment By its RIGHTFUL OWNERS.
Every person meeting with an accident at this Colliery must, before leaving the Colliery premises, report the same to an Official, of the Company, otherwise his claim for Compensation will be resisted.

27.8.1913						
			2 3/4	2	16	10
Coal Cutting					4	6
Clod						
Headings	1	1/10			1	10
Airways	1	1/4½			2	9
Timbers					6	11
Flats	5	1/2				
Cogs						
Rubbish						
Stones	4	4			1	4
Posts In						
Posts Out	1	2/4			2	4
Ripping Top						
Cutting Bottom						
Daywork						
Extra Deadwork						
Allowance						
Total	13	10	...	3	15	2
%	...		...	1	1	0
					13	8
Equivalent	...		...			
Gross Earnings	...		...	5	9	10

DEDUCTIONS.

Coal and Haulage						
Doctor		1	4			
Stores			8			
Levy			8			
Sick Fund			2			
Library Hall and Hospital ...						
Cardiff Infirmary			5			
Advances			9			
Insurance			2			
Checkweigher		1	4			
Sundries			3		5	9
PAY ...				5	4	1

No. 30

Pay ticket of Evan Davies, a miner at Ely Colliery in 1913. It is possible that he had a boy working for him. In this case he would have had to pay the boy from his own wages.

OCEAN COLLIERIES

MAINDY PIT.–UPPER SEAM.

SCHEDULE OF PRICES

Payable for Dead and other Work, when properly and satisfactorily performed.

SEPTEMBER, 1882.

	DESCRIPTION OF WORK.	RATE. s. d.
Cutting ...	Cutting large clean Coal, including Ripping 18 inches of Top	1 11½ per ton.
	Note.—The price paid for the Large Coal covers the price of any Small Coal sent out in trams with the Large Coal.	
Headings	Wide Headings through face of Stalls, single turn	2 9½ per yard.
	Do. do. in solid end of Slips ...	3 9½ "
	Narrow Heading, including Ripping Height to 5ft. 6in. off rail, single turn	5 0 "
	Do. extra double	0 6 "
	Level Headings	1 10 "
	Heading Stall, measuring square off, and Heading Airway in Coal	2 6 "
Timbering	Timbering	1 10 per pair.
	Flat Collars	1 1 "
	Cogs, setting	1 10 each.
	Do. drawing out	1 4 "
	Road Posts, standing (repairing)	0 2½ "
	Posts, drawing out	0 1 "
Ripping .	Ripping and Gobbing Top, per yard, forward, 18in. thick, and not less than 5ft. wide, and in proportion for more or less thickness, but not for extra width	2 3 per yard.
	Cutting Bottom 12in. thick	0 11 "
	Cutting two Ribs in a Stall, when such a Stall is ordered to be driven that necessitates it	0 11 "
Gobbing ...	Tramming Second Rippings, 18in. & upwards when first Rippings was of the usual thickness	3d. to 6d. "

	DESCRIPTION OF WORK.	RATE. s. d.
	Gobbing into Stalls	0 2½ per tram.
Day Work	Gobbing and Labouring	3 0½ per day.
	Timbering, Hitching, Hauling, &c.	3 8 "
Clod ...	Taking down Clod in and across Stall, 9in. thick	Nil.
	Above 9in. and up to 12in. inclusive ...	0 6 per yard
	" 12in. " 15in "	1 10
Working Double ...	Working Double on Coal, 2d. per Ton allowed, but only paid in places urgently required forward.	
Waste ...	Walling Waste, when Walled from bottom to top, leaving about 3 to 4 feet between the face of the Wall and the Rib, and only when a Waste is required to be kept ...	0 6
	Unloading Rubbish in Face of Stalls, including Shifting and Walling in such a manner as to best secure the place, and to the satisfaction of the Officials, on being paid double unloading price, i.e., double 2½d., viz.	0 5 per tram.
	If the Stall is of extraordinary width, owing to special circumstances, then subject to extra payment.	
	Mine (clean), varying according to demand, from	1s. to 2s. 6d. per ton.

LOWER SEAM.

	DESCRIPTION OF WORK	RATE. s. d.
Cutting ...	Cutting large clean Coal, per imperial ton ...	1 5
	Note.—The price paid for the Large Coal covers the price of any Small Coal sent out in Trams with Large Coal.	
Allowance	For Cogs and Walling Sides of Road, see Note in Dead Work Book, September 6th, 1879, only payable when work is properly done	1 $\frac{40}{100}$
Headings.	Wide Heading through Faces, single turn ...	2 7 per yard.
	Do. in end of Slips do	3 1 "
	Do. if Cogged on one side extra ...	0 3 "
	Do. if Cogged on Both sides " ...	0 6 "
	Narrow Headings, single	3 1 "
	Do. extra double	1 0 "
	Level	1 11 "
	Heading Stalls, measuring square off ...	1 5 "
Timbering ...	In Headings, 6ft. 6in. to 7ft. 6in.	1 9 per pair.
	Do. 9ft.	2 0 "
	In Stalls, 6ft. 6in. to 7ft. 6in.	1 4½ "
	Flat Collars	4 1 "
	Cogs, see allowance on Cutting Price of Coal	
	Drawing out Cogs	1 7 each.
	Repairing Posts	0 3 "
	Posts, drawing out	0 1 "
Ripping ...	Ripping and Gobbing Top, per yard, forward, 3(?)in. thick and not less than 5ft. wide, and in proportion for more or less thickness, but not for extra width	3 2 per yard.
Clod ...	Ten inches thick and upwards, in Stalls, when it is requested to be drawn down, but not otherwise (across face)	2 1 "

	DESCRIPTION OF WORK.	RATE. s. d.
Ripping ...	Clod on roadway, 10in. thick, and in like proportion for more or less thickness ...	0 8 per yard.
	Taking down Clod in Headings, not less than 12in. thick	1 0
Bottom ...	Cutting Bottom, 12in. thick	0 11
Gobbing :	Gobbing, per Tram	0 [illegible]
	Labouring and Gobbing	3 0½ per day.
	Tramming and Gobbing Second Rippings, when First Rippings is of the usual thickness, for 18in. thick and upwards, &c. ...	3d. to 6d. per yd.
	Unloading Rubbish in Face of Stalls, including Shifting and Walling in such a manner as to best secure the place, and to the satisfaction of the Officials, on being paid Double Unloading Price, i.e., double 2½d., viz.	0 5 per tram.
	If the Stall is of extraordinary width, owing to special circumstances, then subject to extra payment.	
Day Work	Hitching, Hauling, Timbering, &c.	3 8 per day
	Working Double Time on Coal in Stalls, 2d. per Ton allowed, but only paid in places urgently required forward	0 2 per ton.
	Mine (clean), per Ton, varying according to demand	1s. to 2s. 6d.

WILLIAM JENKINS.

Scale of payment issued to the miners of Maindy Pit in 1882.

Christmas card issued by the coal owners in 1906. Cards like this were sent to coal owners, share owners, directors of collieries, managers and contractors. It was never even considered to send such cards to the ordinary miner to show appreciation for the hard work and dangers experienced during the year. This attitude certainly showed their feeling towards the miners!

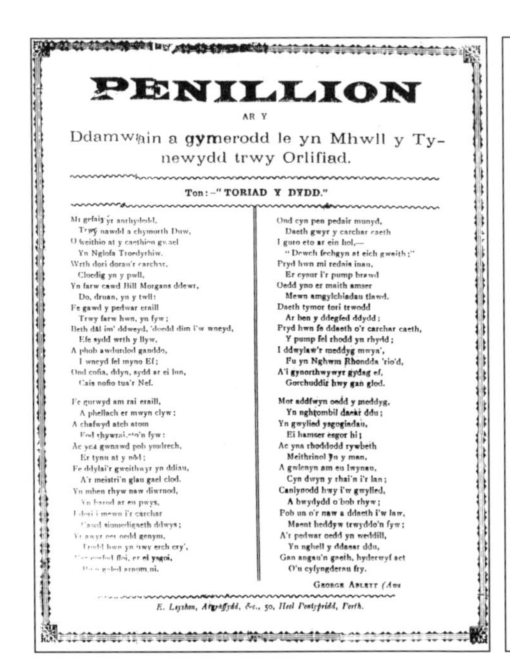

PENILLION

AR Y

Ddamwain a gymerodd le yn Mhwll y Tynewydd trwy Orlifiad.

Ton:—"TORIAD Y DYDD."

Mi gefais yr anrhydedd,
Trwy nawdd a chymorth Duw,
O weithio at y caethion gwael
Yn Nglofa Troedyrhiw.
Wrth dori doran'r carchar,
Cloedig yn y pwll,
Yn farw cawd Bill Morgans ddewr,
Do, druan, yn y twll:
Fe gawd y pedwar eraill
Trwy farw hwn, yn fyw;
Beth dâl im' ddweyd, 'doedd dim i'w wneyd,
Efe sydd wrth y llyw,
A phob awdurdod ganddo,
I wneyd fel myno Ef;
Ond cofia, ddyn, sydd ar ei lun,
Cais nofio tua'r Nef.

Fe gurwyd am rai eraill,
A phellach er mwyn clyw;
A chafwyd ateb atom
Fod rhywrai eto'n fyw:
Ac yna gwnawd pob ymdrech,
Er tynu at y nôd;
Fe ddylai'r gweithwyr yn ddiau,
A'r meistri'n glau gael clod.
Yn mhen rhyw naw diwrnod,
Yn barod at eu pwys,
I dori i mewn i'r carchar
Cawd siomedigaeth ddwys;
Yr awyr oer oedd genym,
[illegible] hwn yn [illegible] erch cry',
[illegible] ffoi, er ei ysgoi,
[illegible] arnom ni.

Ond cyn pen pedair munyd,
Daeth gwyr y carchar caeth
I guro eto ar ein hol,—
"Dewch fechgyn at eich gwaith;"
Pryd hwn mi rednis inau,
Er cysur i'r pump brawd
Oedd yno er maith amser
Mewn amgylchiadau tlawd.
Daeth tymor tori trwodd
Ar ben y ddegfed ddydd;
Pryd hwn fe ddaeth o'r carchar caeth,
Y pump fel rhodd yn rhydd;
I ddwylaw'r meddyg mwya',
Fu yn Nghwm Rhondda 'rio'd,
A'i gynorthwywyr gydag ef,
Gorchuddir hwy gan glod.

Mor addfwyn oedd y meddyg,
Yn nghrombil daear ddu;
Yn gwylied ysgogiadau,
Ei hamser esgor hi;
Ac yna rhoddodd rywbeth
Meithrinol yn y man,
A gwlenyn am eu lwynau,
Cyn dwyn y rhai'n i'r lan;
Canlynodd hwy i'w gwylied,
A bwydydd o bob rhyw;
Pob un o'r naw a ddaeth i'w law,
Maent heddyw trwyddo'n fyw;
A'r pedwar oedd yn weddill,
Yn nghell y ddaear ddu,
Gan angau'n gaeth, hyderwyf aet
O'u cyfyngderau fry.

GEORGE ABLETT (Aw

E. Leyshon, Argraffydd, &c., 50, Heol Pontypridd, Porth.

The Coal Owners' Ten Commandments to the Miners.

I. Thou shalt have no other Master but me.

II. Thou shalt not make for thyself any comforts, nor the likeness of anything - to thine own interest, neither on the earth, nor above, or in the mine.

III. Thou shalt bow down to me, and worship me, for I am thy Master, and a Jealous Master; and I will show you no mercy, but endeavour to make you keep my Commandments.

IV. Thou shalt not take the name of thy Master in vain, lest I sack thee at a minute's notice.

V. Honour thy Master, and his Stewards, and his Deputies, that thy days may be short, and few, for I shall not want thee when thou gettest old and not able to work; thou wilt have to end thy days in the poor-house.

VI. Remember that thou work six days with all thy might and with all thy strength, and do all I want thee, but the seventh day thou stop at home and do no manner of work, but thou shalt do all thou canst to recruit thine exhausted strength for my service on Monday morning.

VII. Thou shalt have no other Union, as it is against my will.

VIII. Thou shalt always speak well of me, though I oppress thee; thou shalt be content if I find thee work and pay thee what I think well; thou shalt starve thyself and thy children if it is anything to my interest; thou must only think of me, not thyself.

IX. Thou shalt have no meetings to consider thine own interest, as I want thee to keep in ignorance and poverty all the days of thy life.

X. Thou shalt not covet thy Master's money, nor his comfort, nor his business, nor anything that is his; thou shalt not grumble at anything, as I want to reign over thee, tyrannise thee, and keep thee in bondage all the days of thy life.

Left: Penillion was written by George Ablett, one of the brave miners involved in the rescue of those entombed at the Tynewydd Colliery, Porth, in 1877, recalling his experiences in the rescue attempts. It was sung to the tune of Toriad y Dydd.

Right: 'The Coal Owners' Ten Commandments to the Miners' were originally distributed to the miners in Yorkshire in October 1920. This copy was then reproduced in *The Colliery Workers Magazine*, Volume 1, No.10, in October 1923.

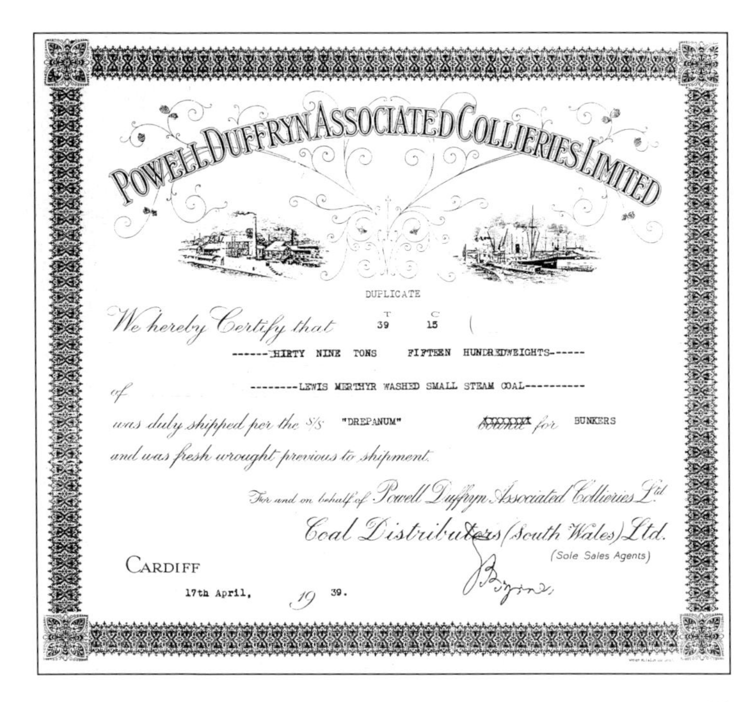

POWELL DUFFRYN ASSOCIATED COLLIERIES LIMITED

DUPLICATE

We hereby Certify that T 39 C 15

------THIRTY NINE TONS FIFTEEN HUNDREDWEIGHTS------

of --------LEWIS MERTHYR WASHED SMALL STEAM COAL----------

was duly shipped per the S/S "DREPANUM" bound for BUNKERS

and was fresh wrought previous to shipment.

For and on behalf of Powell Duffryn Associated Collieries Ltd
Coal Distributors (South Wales) Ltd.
(Sole Sales Agents)

CARDIFF

17th April, 19 39.

A typical Export Certificate for coal exported from Cardiff Docks. The date is 17 April 1939.

'Don't Go Down The Mine Dad
A miner was leaving his home for work, When he heard his little child scream; He went to his bedside, his little white face, Oh, Daddy, I've had such a dream: I dreamt that I saw the pit all afire, And men struggled hard for their lives; The scene it then changed, and the top of the mine Was surrounded by sweethearts and wives.'

'Don't go down the mine, Dad, Dreams very often come true; Daddy, you know it would break my heart, If anything happened to you; Just go and tell my dream to your mates, And as true as the stars that shine, Something is going to happen today, Dear Daddy, don't go down the mine!'

'The miner, a man with a heart good and kind, Sat by the side of his son; He said: Its my living, I can't stay away, For duty, my lad, must be done. The little one lookd up, and sadly he said: Oh, please stay to-day with me, Dad! But as the brave miner went forth to his work, He heard this appeal from his lad.'

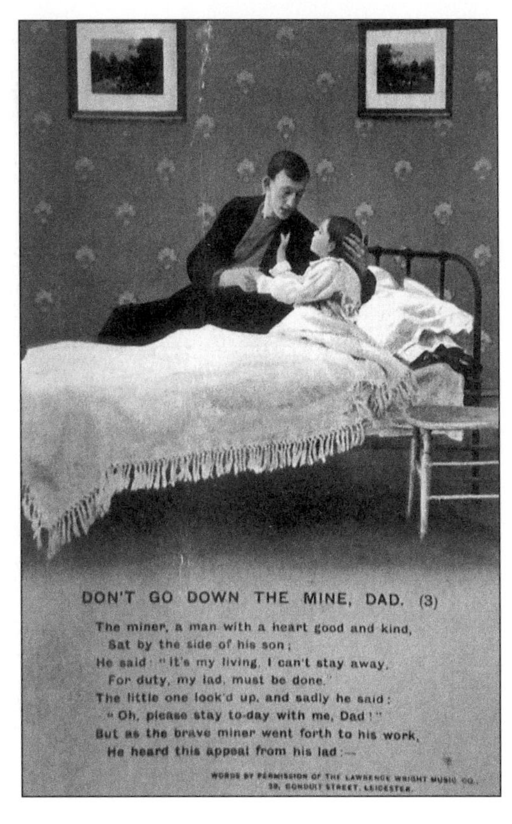

'Whilst waiting his turn with his mates to descend He could not banish his fears; He returned home again to his wife and his child, Those words seemd to ring through his ears. And ere the day ended the pit was on fire, When a score of brave men lost their lives; He thanked God above for the dream his child had, As once more the little one cried: – Don't Go Down The Mine Dad.'

These songs were sung in the local Music Halls in the early part of the 1900s and were very popular with the miners and their families who visited them. They were later printed on to this set of four postcards by a company called Bamforth. This company had obtained permission to print these words from the Lawrence Wright Music Company, of Leicester.

Typical levels tunnelled into the mountains in the 1900s. (Photograph by courtesy of Mr Glyn Gregory, Clydach Vale).

During the strike years of 1910 and 1911, the miners would scavenge for coal from the tips. Younger and more able miners would dig into the mountainsides in search of coal. The levels produced much needed coal for the men and their families. Evidence of such levels can still be found on the mountainsides of the Rhondda today. (Photograph by courtesy of Mr Glyn Gregory, Clydach Vale).

Two

Working and Living Conditions

In the early 1800s both the Rhondda Fawr and the Fach were inhabited by isolated farms, sparsely distributed throughout the area. Prior to 1860, when the presence of coal in the lower part of the Rhondda Fawr was first discovered, housing development had only started in certain areas of Porth and Cymmer. These early housing developments were often in isolated, single-storey rows of cottages built either near the pit-head of the local colliery or alongside the old parish roads which ran throughout the valley. Depending on their location, they sometimes had gardens at the front of the row. More often small gardens were located at the rear of each row of terraces.

Due to the geographical formation of the valleys, the valley floor was used for the working and distribution of coal, with later housing in the form of terraces built on the sides of the valley, in line with the valley floor and the river which flowed there. This became the most economical method of utilizing the land available, resulting in the row after row of terraces we see today.

When considering that, in 1861, the population of the valleys was approximately 4,000 and had risen to approximately 163,000 by 1921, the number of houses rising accordingly from 561 to over 29,000 within the same period, one can appreciate the problem of providing housing at this time. All the available space between the pit-head settlements were filled with more rows of terraces, resulting in a continuous chain from the mouth of the valley to its head, punctuated by the collieries which provided work for all. However, the building of these rows of cottages created other problems: sanitation had a lot to be desired in these early days.

Although there was employment, wages were not high. Families were usually of four or five, all living in small cottages. In order to supplement their low income, it was usual for the family to take in a lodger resulting in overcrowded conditions. Getting clean water was nearly impossible and this, together with the primitive sanitary systems of the time, ensured that diseases such as tuberculosis, rickets, and malnutrition were common, especially among children, resulting in appalling rates of infant mortality. These conditions and the loss of life in the mines helped to foster and develop the community spirit in which the Welsh valleys and especially the Rhondda Valleys were well known and in those days helped the people to at least come to terms with their problems, even if they did not solve them.

With greater demand from the coal owners for more coal output and with shortcomings on safety within the collieries, it was not long before accidents started to take place. It would not be long before a major disaster would occur.

In the Rhondda, fifteen major explosions took place over a period of sixty years resulting in the deaths of approximately 850 men and boys. In some explosions whole families were wiped out in one sweep. In other instances local communities, who were just coming to terms with earlier disasters, were once again subjected to further agony. During this period there were not many houses in the valleys which had not suffered from one tragedy or another. How did those left behind manage? Due to the low wages paid by the coal owners, there was little money left for essentials, let alone payments to a Friendly Society or an Insurance Company. In the event of another tragedy, people had to rely on the generosity of relief funds, usually set up very quickly by local organizations, including local businesses and individuals, who gave what they could to help those in need. The coal owners, living at the height of luxury, were very quick at the time of such disasters to contribute to such funds with the maximum publicity on their generosity! Such an example of their generosity is shown for the terrible explosion at Mardy Colliery on Wednesday 23 December 1885, when eighty-one men and boys were killed. Locketts Merthyr Steam Coal Co. gave £300; Glamorgan Coal Co. gave £50; Lord Bute £200 and Lord Aberdare £50.

It was a pity that more money was not readily available to improve safety standards within their own collieries, so that tragedies such as this might have been avoided. Those families which had received extra money from such donations had their hardship relieved for a short while but they, like others who had suffered, did not have the means to protect themselves from the future. They could not claim any form of compensation from the coal owners, as records show without exception that at the inquest of any major disaster, the verdict was accidental death. Therefore, no money could be claimed from the coal owners to provide for the future of the miners and their families.

In order to survive, the bereaved widows applied for Parish Relief which was means tested. If it was felt that the family had any assets of value, such as furniture or even ornaments, these had to be sold and the money used to clothe and feed the families. It was only when they had nothing more to sell and were reduced to the lowest form of poverty, that any monetary relief was given. This system carried a stigma with it and many families were loath to apply but did so out of pure desperation. This type of situation was experienced for many years until the families' children were old enough to work and contribute to the family income. The money from the Parish Relief was immediately reduced but at least the family had attained some independence from a system designed to strip people of their dignity.

In the early twentieth century the miners organized themselves into a Union. Its main aim was to improve the minimum wage and the conditions in which miners worked. It also set up a financial programme to aid the widows of miners killed whilst working at the collieries. It would, however, be many years before any change was seen.

At the beginning of the First World War, in 1914, there was a remarkable increase in the demand for coal to fire the ships' boilers. Since Rhondda Steam coal was ideal for this, the miners had high hopes that, with the nationalization of the mines, wages and conditions would improve and continue to improve. However it was not to be.

Following the end of the First World War, the mines were returned to the coal owners who set about lowering the miners wages, arguing that the government had been too generous in their payments during the war years. The 1920s saw a new era of bitterness between the miners and the coal owners.

In order to enforce the new wage levels, the owners gave the miners an ultimatum: accept the new pay structure, or offers of employment would be withdrawn. The miners refused to accept this ultimatum, with the result of a lockout throughout the coalfields of Great Britain, with over one million miners being unemployed. This dispute lasted for three months. In the end the miners were forced to return to work and accept the reduced wage scales. During the following years, a conditions and wages improved slightly in the valleys as the demand for coal grew to meet the demands of Industry and Commerce.

However, to meet the demands of dwindling profits, pay cuts were demanded by the coal owners. Again a bitter dispute arose between the miners and the owners and, with support being promised by the newly formed Trade Union Congress calling for a national strike, the pits closed. Moreover, within days the Trade Union Congress withdrew their support and the miners were left to face the consequence of their actions alone. The 1926 strike was very bitter but, due to the poverty experienced throughout the South Wales Coalfield, by the end of the year even the most hardened of the men were forced back to work. Even today, amongst families of future generations they talk of the hardship experienced through this bitter strike!

The outbreak of the Second World War saw improvements in the conditions of the mineworkers and Nationalization in 1947 brought the promise of prosperity for the coalfield. On 1 January vesting day celebrations took place in the Rhondda Valleys with a series of concerts, brass bands, mannequin parades and other social activities.

Unfortunately, these promises were not going to be fulfilled: nationalization did little to tackle the problems of the coal-mining industry which was already in decline. Within a few short years pit closures began to spread throughout the South Wales Coalfield. From the mid-1950s, pit closures and the consequent loss of unemployment accelerated throughout the industry, while, in the 1960s, no less than seventy-four collieries were closed across the coalfield.

During this period in the Rhondda many of the collieries were closed and after Lewis Merthyr closed in 1983, Mardy was the only colliery producing coal in the Rhondda Valley! This colliery closed in December 1990 and was known as the last working pit in the Rhondda.

In October 1993 a bronze statue of a miner with his wife and baby in her arms, commemorating Rhondda's Past was unveiled on the site of the Scotch Colliery by Viscount Tonypandy in the presence of many honoured guests and local dignitaries. Eight-feet tall and standing on a plinth of the same height, the couple, with their baby in a shawl nursed-Welsh fashion, will be a powerful reminder of Rhondda mining history. The cost of the memorial, sponsored by the Rhondda Civic Society, was in the region of £70,000. The sculptor, Robert Thomas, originally of Cwmparc, resided in Barry until his death in May 1999.

Three
Coal Owners in the Rhondda

George Insole

George Insole was born in Worcester in 1790. Son of cabinet maker William Insole, he was apprenticed to his father and became a cabinet maker in his own right. He married Mary Finch in August 1819 and his eldest son James Harvy was born in 1821.

In 1827 he moved to Cardiff with his family and set up a company in partnership with Richard Biddle, their main business at that time being the trade in timber. In 1828, however, he turned to the exporting of coal and became the chief shipping agent for Robert and Lucy Thomas, farmers turned coal owners, of Abercanaid, near Merthyr. Between 1828 and 1850 he became a market specialist in the exporting of Welsh steam coal to the London markets.

Being a shrewd businessman, aware of the increasing demand for this type of coal, in 1832 Insole leased the mineral rights of the Maesbach Estate in Llantwit Fadre, and sunk the pit called the Maesbach Colliery. He soon realised that the supply from this colliery could not meet the demands of the London and Irish markets and looked for ways in which to expand his company's potential.

By 1844, together with his son James Harvey Insole, he secured the mineral rights of 375 acres of Cymmer land from Evan Morgan of Tyn Y Cymmer Farm. In 1847 he sunk the Cymmer No.1 pit to work the No.3 Rhondda seam. However, in order to meet the increasing demand for coal, he further extended mining operations at Cymmer by sinking the Upper Cymmer Colliery in 1851, the year he died.

The New Cymmer Colliery was sunk in 1855 by his son who, in 1875, decided to renew attempts to reach the lower and richer coal seams, achieving this in 1877. Whilst other collieries in the area merged into different groups of Companies, Insole's Cymmer Collieries was one of the few individual enterprises remaining in the Rhondda Coalfields in 1929 and was registered as Messrs. Insoles Limited.

Whilst a few of the larger coal owners lived in the area and showed an interest in the welfare and education of their employees, the majority unfortunately did not. George Insole was one of these, preferring to spend his money on a lavish style of living, which he enjoyed at his large house, The Court, Llandaff, Cardiff.

Insole and his son only visited the Cymmer Colliery weekly, leaving the everyday running of the colliery under the direction of his Agent and Manager, Jabez Thomas.

Walter Coffin

Walter Coffin was born in Bridgend in 1785, the second son of Walter Coffin Senr, the founder of a large and successful leather and tanning business in the area. Educated at Cowbridge Grammar School and later at a private school (Academy) in Exeter, he was encouraged to enter the family business but showed little interests. Well aware of the potential of the expanding South Wales coalfield and of the high demand of coal, he bought a farm at Dinas in 1809 and began prospecting for coal, opening a level on the farm property.

In 1810 he took on a lease from Morgan David for the mineral rights under the Graigddu and Gwaunadda farms for ninety-nine years at a rent of £30 per annum, and sufficient coal for the lessor's personal use. In 1812, he sank his first pit, the Dinas Lower Colliery, on the land leased to him. This pit, the first to be sunk in the Rhondda Valleys, struck an excellent seam of coal at a depth of forty yards. Marketed by him as Dinas No.3, the seam became widely known as Coffin's Coal. After the death of his father, also in 1812, the family gave up the home at Nolton Bridgend and moved to Llandaff Court, Cardiff. Coffin became Mayor of Cardiff in 1848 and, in 1852, Member of Parliament, a position he held until 1857.

By this stage, having accrued a large fortune and established himself as a major player in South Wales politics, his interest in the South Wales coalfield had diminished. After ceasing to be a Member of Parliament, he spent the latter years of his life at his home in Princess Gate, Mayfair London, where he died in 1867, aged eighty-two. As neither he or the other members of the family had married, the Coffin family became extinct, and his estate was divided among distant relatives and close friends.

David Alfred Thomas

David Alfred Thomas, was born at Yscuborwen Aberdare, in 1856, one of five children. Educated at Clifton School and at Cambridge, he was a brilliant mathematician, with a real interest in other scientific subjects, but cared little for literature and the classics.

He suffered ill health, having rheumatic fever when quite young. However, in spite of his delicate health, he was quite a keen sportsman. After the death of his father Samuel Thomas, he left Cambridge in 1890 and went to Clydach Vale where he lived for a short time in order to gain experience in the mining industry. Thomas' father had sunk the Clydach Vale Colliery in 1874.

In 1882 he married Miss Sybil Margaret Haig and had one daughter by the marriage. At this time he was residing in London, where he entered a stockbroker's office only to return to Cardiff after a short period of time. In due course, through his many other undertakings and directorships, Thomas became a prominent member of the South Wales Coal Owners' and Export fraternity.

In 1887 he took on a lease of Llanwern House, Nr. Newport, where he resided for the remainder of his life. In 1888 he became the Liberal Member of Parliament for Merthyr,. He remained a member until 1910, when, for a period of six years, he represented the constituency of Cardiff. During the First World War Thomas was employed in government services and, in 1915, he went to America on behalf of the Ministry of Munitions. In 1916 he was appointed Food Minister and following year was created a Viscount, taking the title of Lord Rhondda.

Although he was heavily involved in politics, Thomas had returned in 1906, to Clydach Vale, devoting his energies and talents to the development of the Cambrian Collieries. By 1908 the Cambrian Combine, then most important amalgamation of a group of collieries in the Rhondda Valley was formed. He died in 1918 aged sixty-two years, at his home in Newport, whilst still Managing Director of the Cambrian Combine Collieries.

James (Siamps) Thomas

James (Siamps) Thomas was born on a farm in Bedwellty in 1817 and, from the early age of six years, worked as a door-boy in a local colliery. He became a fireman, overman and eventually worked his way up to the position of pit manager. From 1840 he occupied various responsible positions in collieries in Glamorganshire and Monmouthshire. A shrewd businessman, by 1850 Thomas was in a position to purchase a share in the partnership of the Troedyrhiw (Aberhondda) Colliery, Porth. He quickly became the joint owner of the Ynysfeio Colliery, at Treherbert, and the Tynewydd Colliery at Porth.

During 1877 tragedy struck at the Tynewydd Colliery in Porth with the flooding of the mine workings, resulting in the deaths of five miners. Five other miners were rescued after they had been trapped in an air pocket for nearly ten days. After the drama of the rescue had receded, a coroner's inquest was opened. The verdict was that James Thomas, Manager of the colliery, was found guilty of manslaughter. The case was then referred to the Glamorgan Summer Assizes at Swansea in August 1877. The trial, which lasted two days ended with the jury, made up of shopkeepers, farmers, business people, etc; not being able to agree on a verdict, pleading lack of specialist geological knowledge. The case was reopened at Cardiff Spring Assizes in April 1878 and, after a short trial, the jury returned an unanimous verdict of not guilty. The aquittal was greeted with applause which was speedily suppressed. (For further details of this trial refer to the book Disaster at Tynewydd by Ken Llewellyn.)

Lacking in education, Thomas took no part in public life, and had little interest outside the affairs of his collieries. In later life, having given up his active role in the management of his collieries, Thomas turned to agriculture and farming, and purchased a good deal of land in the Counties of Glamorgan and Monmouth. He died of pneumonia in 1901 at his home in Cross Farm, Llanedarne, near St Mellons, at the age of eighty-four, leaving behind a son and three daughters.

The family tomb and the statue of James (Siamps) Thomas in the churchyard of New Bethel Chapel, at Mynydd Islwyn, Gwent.

Archibald Hood

Archibald Hood was born in June 1823 at Kilmarnoch, Ayrshire and, as a youngster, only received a basic education. By the age of thirteen, he was working at a local colliery, being in charge of a surface engine and working approximately twelve hours a day. When he was about seventeen years old, he had the opportunity to become involved in further study, eventually becoming a qualified mining engineer.

In 1860, he joined the Ely Valley Coal Company at Tonyrefail and, to safeguard his business interests in the area, moved from Scotland to Gilfach Goch, residing at Gilfach House. He then proceeded to negotiate mineral rights with landowners in the area and, in March 1863, began sinking for coal at Llwynypia. Hood changed the name of the Company from Ely Valley Coal Company to the Glamorgan Coal Company. By March 1864, he had struck the No.2 coal seam and, in January 1865, the No.3 coal seam was reached.

In 1867, due to the expansion of the Glamorgan Coal Company, he took up residence with his family at Sherwood Newport Road, Cardiff, although retaining his Scottish home, Rosedale House, at Rosewell, Midlothian. He died at his house in Cardiff in October 1902, at the age of seventy-nine and was buried at Cardiff Cemetery.

After the first shafts were sunk in 1863, more workers migrated to the valleys. Hood built the Llwynypia Terraces to house his workers. Because of the Scottish influence, they quickly became known as the Scotch Terraces and are still known by this name today. They were built in a different style to the other miners cottages, as each cottage had a large garden attached to the it which enabled the miners to grow vegetables for the benefit of themselves and their families.

Hood also introduced a system of education for the sons and daughters of his workers and built a Workmens' Institute with library facilities, newsrooms and even swimming baths for sporting activities.

Due to the high esteem in which he was held by his workers, it was decided to commission a fitting memorial to his memory. To this effect a life size bronze statue of Hood, pointing towards his colliery, was erected in the grounds of the Llwynypia Library and Institute and unveiled on Federation Day in July 1906 by William Abraham (Mabon) MP. This statue was paid for by the workers of his collieries who had contributed to the erection of the statue, and had raised over £600 for it (a very large sum of money in those days). This was the first public statue to be erected in the Rhondda.

The statue of Archibold Hood located on its foundation, prior to its official unveiling by Mr W. Abrahm (Mabon) MP on Federation Day, July 1906.

Moses Rowlands

Moses Rowlands was the son of Moses Rowlands Senr, who was invited by Walter Coffin to come from Carmarthanshire in 1825 in order to teach Arithmetic and English to the children of his miners in Dinas and Penygraig. He became such a firm favourite with Coffin that he later became his secretary and his Clerk of Works. He died in 1837.

His son, Moses Rowlands Jnr, was a qualified mining engineer and, in partnership with a friend, the geologist Richard Jenkins, opened the Penygraig Level in 1857 and worked the outcrop of the No.2 Rhondda seam. Realizing the potential of mineral resources in the area and seeing that they were not being exploited by other old Rhondda families, he formed a partnership with William Morgan, William Williams, and John Crockett, all from Pontypridd, to form the Penygraig Coal Company. In 1864 they sank a small pit known as the Penygraig Colliery to the Nos.2 & 3 Rhondda seams, an extremely successful venture, which by 1870 was producing an annual output of approximately 100,000 tons. This colliery continued successfully until 1919, when the lease expired.

In about 1875, Moses Rowlands and William Morgan, determined to take advantage of the high market price of good quality steam coal of the time, founded the Naval Colliery Company. They opened the Pandy Pit and reached the main coal seams in 1879, where over 76,000 tons of good quality steam coal was raised the following year.

However, in later years, geological problems were encountered. Two explosions occurred in 1880, with the death of ninety-six miners and in 1884, when fourteen miners and officials also perished.

Moses Rowlands died at his home in Penygraig in 1885 and the company was sold in 1887. It was later renamed the New Naval Collieries Company.

Alfred Taylor

Alfred Taylor was born in 1824 at Burnshill Row, Shoreditch. He came from a Quaker family, and was educated at schools belonging to the Society of Friends. A well known geologist and a member of the Geological Society, Taylor was also a metallurgist, his knowledge and experience gained through working in the family foundry business.

In 1872 Taylor bought the mineral rights of Penrhys Farm and subsequently sunk two shafts, known as the Perhys Colliery. Efforts to accommodate the miners who came to work at this colliery resulted in the development of the township of Tylorstown, named after him. It would appear that Alfred Taylor did not spend a great deal of his time on this new enterprise as, in 1876, his brother Louis Taylor, who was a partner in this venture, gave up his business interests in Yorkshire to take charge of the collieries in South Wales.

He died in Carshalton in Surrey in 1884, at the age of sixty. His brother Louis died in Redhill Surrey in 1905.

David Davies

David Davies was born in December 1818 at a farmhouse at Llandinham, Montgomeryshire, (absorbed into the new county of Powys in the government reorganization of 1974). The eldest of nine children, during his childhood Davies worked on the farm which was nothing more than a small holding. It was mostly through his efforts that the family were able to move to a larger farm, situated on the valley floor, in Llandinham called Neuaddfach. Determined to succeed in life, he took small contracting jobs for the County Highways Surveyor, Thomas Penson, repairing roads and bridges. He did the work so well that Penson gave him the contract for laying the foundations of a new bridge to be built over the river Severn, at Llandinham. This was the first cast iron bridge to be built in the county and is still in use today.

In 1851, he married Margaret Jones of Llanfair Caereinion and their only child, Edward, was born the following year. In 1885 he built the first railway in Mid-Wales, between Llanidloes and Newtown. Other railways throughout Wales, such as the Shrewsbury to Aberystwyth sections between Newtown and Machynlleth, were built by him. In later years he became known as Railway King of Wales.

Having made his fortune, Davies built his new home, a mansion called Broneiron, on a prominent hillside in Llandinham. In 1864, he diverted some of his energies from railway construction and came to the Rhondda Valley to exploit the mineral resources of steam coal which was in great demand as steamers were taking the place of sailing ships all over the world. He negotiated leases at Ton-Pentre and Cwmparc. After months of uncertainty, with the shaft being sunk deeper and deeper, and him about to give up, the miners volunteered to work for a week without pay. During that week in March 1866 at a depth of 215 yards, one of the finest seams of steam coal in the South Wales Coalfield was struck. Being such an astute businessman, Davies marketed this coal under the title of Ocean Merthyr coal, few people at that time had heard of the Rhondda, but the town of Merthyr was known the world over!

In 1874 he became Liberal Member of Parliament but he never had great success in politics and eventually lost his seat in 1886. His collieries continued to expand. At the time of his death, he was employing 5,000 men, and raising 1,600,000 tons of coal per annum. The rate of output growth was so rapid that the docks in Cardiff could not cope. David Davies, along with other Rhondda Coal owners were instrumental in creating the docks at Barry. The docks were officially opened in July 1889, at a cost of £2 million. The chairman of the newly-formed Barry Dock and Railway Company was the Earl of Plymouth, while David Davies was appointed vice chairman. Davies worked on developing the docks up until his death in 1890, at the age of seventy-two. His bronze statue, the work of Sir Alfred Gilbert, sculptor of Eros in Picadilly Circus, London, still stands outside the imposing Barry Dock Offices. He was buried at Llandinham. A replica bronze statue of him stands near the first cast iron bridge where he had laid the foundations besides the main road through Llandinham.

The statue of David Davies of Llandinham located outside the Barry Dock Offices.

Edmund (Hanney) Watts

Edmund (Hanney) Watts was born in 1830 at Blyth, Northumberland, and, at the age of fifteen, was apprenticed to a Newcastle Company called Balderman, Bariers, & Co.

At the age of twenty-one, after completing his apprenticeship, he returned to his native town of Blyth and set up in business as a merchant and shipbroker. His business thrived and consequently he became one of the leading ship owners in the country. In 1872 he expanded his operations by purchasing the Tydraw Colliery at Blaencwm and, through his shipping interests expanded, his business as a coal exporter.

In the early 1880s he sunk two collieries in the area which was later called Wattstown after him. Although he never lived in the area, he made frequent visits and was quite involved with the commercial life of the district. He was also one of the original promoters of the development of the Barry Dock and Railway undertaking.

He died in 1902, aged seventy-two at his home at St Leonards on Sea, where he is buried.

Four

The Rhondda Collieries

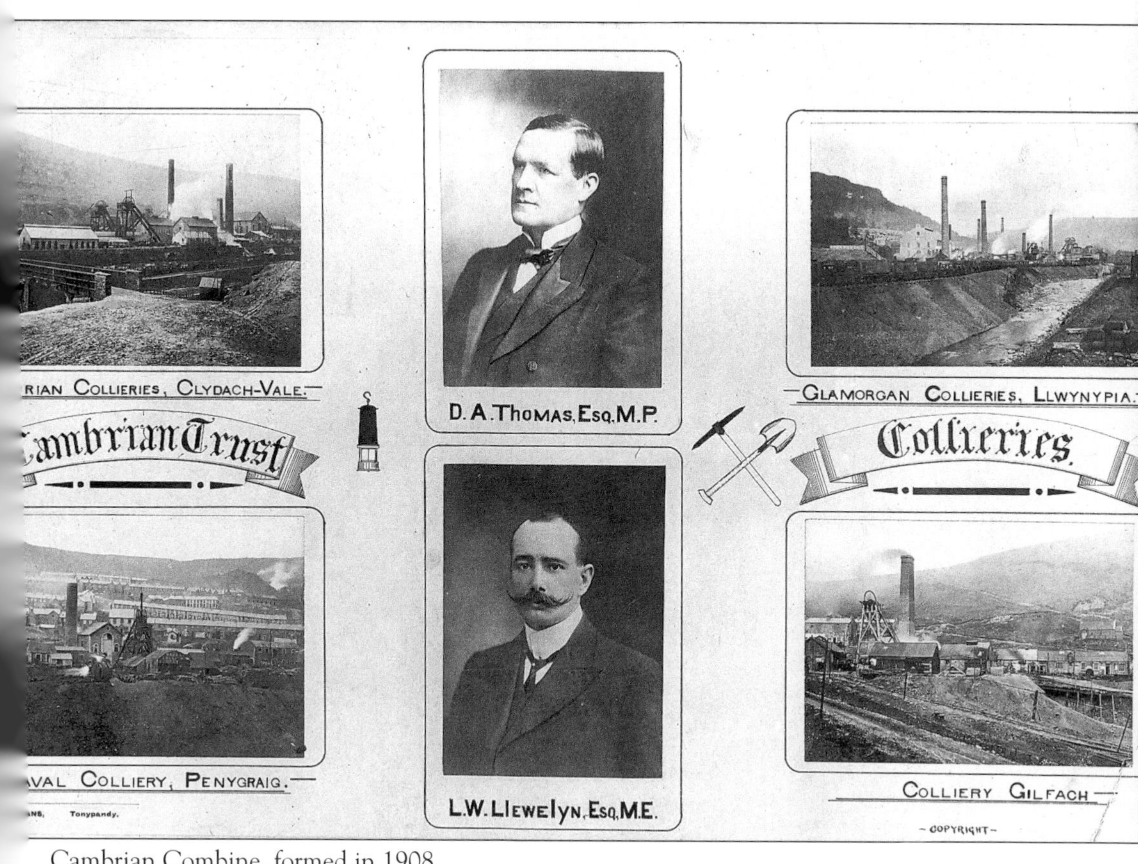

Cambrian Combine, formed in 1908.

Colliery closures

Here is a list of locations, collieries' descriptions and date of closure:

Trehafod
Lewis Merthyr, 1983

Porth,
Llwyncelyn, 1895
Cymmer, 1940
Tynewydd, 1901

Rhondda Fawr

Dinas
Lower & Middle, 1893

Penygraig
Ely, 1958

Tonypandy
Nantgwyn, Pandy (Naval), 1958
Anthony, 1952

Clydach Vale
Blaenclydach, 1947
Cambrian, 1966

Llwynypia
Glamorgan (Scotch), 1945

Pentre
Gelli, 1962
Bodringallt, 1959
Maindy, 1948
Pentre, 1929
Tynybedw, 1933

Cwmparc
Dare, 1966
Parc, 1966

Treorchy
Tylecoch, 1895
Abergorki, 1938

Treherbert
Ynysfaio, 1947
Bute, 1926
Lady Margaret, 1909
Tynewydd, 1911

Blaencwm
Tydraw, 1959
Glenrhondda (Hook & Eye), 1966

Blaenrhondda
Blaenrhondda, Fernhill, 1966

Rhondda Fach

Ynyshir
Ynyshir, 1909
Lady Lewis, 1926
Standard, 1947

Wattstown
National, 1968

Tylorstown
Tylorstown Nos6, 7, 8, 1936
Tylorstown No.9, 1960

Ferndale
Ferndale No.1, 1959
Ferndale Nos2, 4, 1930s
Ferndale No.5, 1959

Maerdy
Mardy Nosl & 2, 1932
Mardy Nos3 & 4, 1990

Mining Disasters in the Rhondda Valleys

The following are the date of the disasters, the pits and the number of people who were killed:

1817 Abergorchy level, 1
1836 Dinas Lower, 4
1939 Dinas Colliery, 3
1844 Dinas Middle, 12
1844 Fforest Level, 6
1856 Cymmer Colliery, 114
1857 Ynyshir Colliery, 4
1857 Cymmer Old, 2
1862 Ynysfeio Colliery, 2
1865 Bute Colliery, 4
1866 Tylacoch Colliery, 2
1867 Ferndale Colliery, 178
1868 Tylacoch Colliery, 2
1869 Ferndale Colliery, 53
1871 Pentre Colliery, 38
1871 Gelli Colliery, 4
1875 Naval Colliery, 2
1877 Tynewydd Colliery, 5
1878 Dunraven Colliery, 2
1879 Dinas Colliery, 63
1880 Naval Colliery, 101
1882 Parc Colliery, 4
1882 Coedcae Colliery, 6
1883 Coedcae Colliery, 5
1883 Gelli Colliery, 5
1884 Naval Colliery, 14
1885 Maerdy Colliery, 81
1887 National Colliery, 39
1892 Great Western Trehafod, 58
1896 Tylorstown Colliery, 57
1901 Dinas Level, 4
1903 Ferndale No.5 Colliery, 1
1904 Llwynypia No.3 Colliery, 1
1905 Cambrian Colliery, 33
1905 Tylorstown Colliery, 2
1905 National No.2 Colliery, 119
1905 Ferndale No.9 Colliery, 2
1906 Penygraig Colliery, 2
1932 Llwynypia Colliery, 11
1941 Gorki Drift Mine, 7
1956 Lewis Merthyr Colliery, 9
1965 Cambrian Colliery, 31.

Trehafod (Summer House)

A view of the Lewis Merthyr Collieries, Porth, in a postcard from the 1910s.

The Lewis Merthyr Collieries.

In 1850 brothers David and John Thomas leased the mineral rights of Nyth-Bran and Hafod Fach farms, on the eastern bank of the river Rhondda and sank the Hafod shaft to work the No.3 seam but soon encountered difficulties. The brothers deepened the shaft in order to work the Hafod seam but, after spending over £30,000 they had to abandon the project because of geological problems.

In the same year, but on the south bank of the river, Edward Mills sank the Coedcae Colliery. He soon encountered water problems as well and the site was abandoned. The Thomas brothers later tried to rework this colliery but also experienced water problems and abandoned the project.

In the mid 1870s William Thomas Lewis (later to become Lord Merthyr) purchased both collieries. By 1880 W.T. Lewis had sunk the Bertie shaft and, in 1890, the Trefor shaft; both shafts were named after his sons. In 1881 Lewis Merthyr Navigation Collieries Limited was formed. By 1900, they were consolidated under the single control of the Lewis Merthyr Consolidated Collieries Limited, which by this time was producing a million tons of coal annually. Before the First World War the output of the company rose to approximately 1,500,000 tons per year.

In 1929 the Powell Duffryn Company took control and, in the same year, Coedcae colliery stopped raising coal, the shaft being retained for pumping purposes only. The year 1935 saw the amalgamation of Powell Duffryn and the Welsh Associated Collieries, which became one of the largest coal producing groups in the world. By 1937, approximately 1,300 miners were working at Lewis Merthyr. At the time of the explosion in November 1956 that killed nine men, the number of miners employed had been reduced to approximately 1,150.

In 1958 Lewis Merthyr and Tymawr collieries were merged. From the time of the merger no coal was raised at Lewis Merthyr, all coal produced being raised at the Tymawr pithead. In July 1983 both collieries were closed.

Lewis Merthyr Colliery, Porth, 22 November 1956.
On 22 November 1956, gas, which had accumulated in a roadway cavity, was ignited by a fall of stone. At that time about thirty miners were working in this district of the pit. Two of the miners were killed immediately while twelve others were injured. Later seven of those who were badly burnt died from their injuries, bringing the total number of deaths to nine.

The names of those who died are:

Arthur Atkins, 40 years, Collier, Ynyshir
Trevor Davies, 38 years, Charge Hand, Porth
Albert R. Fox, 41 years, Colliery Manager, Porth
Elias Howells, 37 years, Turbine Attendant, Porth
Gildas Jones, 36 years, Overman, Porth
Richard Jones, 57 years, Repairer, Porth
John Miles, 35 years, Ass. Repairer, Penygraig
Peter Proffitt, 27 years, Ass. Repairer, Cardiff
Sylvanus F. Thomas, 69 years, Overman, Trehafod

The names of those who were injured are:

H. Bryant, 46 years, Chargeman
W. Childs, 43 years, Repairer
F. Crump, 39 years, Repairer
W.H. Davies, 50 years, Packer
I. Humphries, 35 years, Deputy

The Lewis Merthyr Collieries, c.1912.

The colliery area today. Note the new Trehafod by-pass road.

The Rhondda Heritage Park.

When the Lewis Merthyr Colliery closed in 1983, a body of local people came together to ensure that their heritage and the spirit of the coal mining communities of the Rhondda Valleys was preserved for future generations. The development began with the restoration of the main colliery buildings, with painstaking attention to detail, the clearing away of unwanted buildings and obsolete machinery and with everything restored to its former glory. The visitor centre opened to the public in 1989. The first major exhibition of Black Gold opened in 1991 and relates to the unique character and culture of the Rhondda Valleys seen through the eyes of three generations of one local mining family.

Further developments took place with the opening, in 1993, of the Trefor and Bertie Energy Zone, an action-packed play area based on energy cycle. Here children are able to discover how coal was created and are involved in various exhibits where they are able to mine their own coal. In July 1994, a new development was completed. A simulated underground tour called A Shift In Time was opened to the public. This complemented the Black Gold Exhibition and the Energy Zone complex, thus completing the development programme at the Rhondda Heritage Park.

The tour guides employed at the Park are all ex-miners, with personal knowledge and experience of the hardships, tragedies and humour shared by the miners and their communities. During the tours, they are able to share their experiences with the visitors.

A permanent exhibition relating to the famous Tynewydd Colliery Disaster, which took place at Porth in 1877, also opened on 21 February 1999. The impressive murals were painted by the Trehafod artist Richard Oliver. This latest development ensures that the updating of historical mining events which took place in the Rhondda continues. The number of visitors to the Park is increasing, thus ensuring its future. The park is now the flagship of a rapidly developing tourism industry in the Rhondda.

The original Trefor and Bertie winding houses, incorporated into the Heritage Park complex.

The interior showing items of interest such as chapel seats from local churches, Salvation Army drums and metal signs advertising different products.

A view of the Trefor Winding House and Pit Gear. The drams shown are those that were used to raise the coal from the pit bottom.

Porth (The Gateway)

Llwyncelyn Colliery, Porth

Sunk in 1851 by David Jones of Merthyr, the Llwyncelyn Colliery was owned and worked by T. Edmunds who also operated a set of coke ovens on the site, whilst working the No.2 and No.3 Rhondda coal seams.

In 1870 Edmund Thomas purchased this colliery from Christopher Jones and Company, the owners at this time. However, many problems were encountered in working the seams and the colliery was subsequently closed in 1883.

In 1891 the colliery was reopened under the ownership of the Lewis Merthyr Colliery Company. Ongoing geological problems forced further closure in 1895. The site was then used by the company for many years as a storage depot. In later years, the property was leased by the local Electricity Board which installed a generating plant to supply the area and also for the storage of their equipment. Today the site has been cleared and is used by a local garage for storage, maintenance and repairs of vehicles.

Above and right: Llwyncelyn Collieries, c.1909.

The site of the Llwyncelyn Colliery today. The capped shaft can be seen at the bottom right hand corner.

The Britannia Inn. Attached to the end of a row of cottages, this public house appears to have been used by the miners, not only as a local watering hole, but also to collect their wages. The Britannia Inn has now been renamed The Lodge. Although no longer used by the miners, it is still a popular venue with the local community.

Cymmer Colliery, Porth.

George Insole started in the coal trade as the Chief Agent at Cardiff for farmers turned coal owners, Robert and Lucy Thomas from Abercanaid, Nr. Merthyr, who were the first coal owners to ship coal to London, in 1828.

George Insole, who owned a shipping office in Cardiff since 1828, had for several years been a market specialist in the exporting of coal. The high demand for Welsh steam coal on the London markets prompted him to become a producer as well as a marketer. In 1844, Insole leased the mineral rights of 375 acres of Cymmer land from Evan Morgan of Tynycymmer Farm. Although he worked the Rhondda No.2 seam through various levels, it wasn't until 1847 that he sank the Cymmer No.1 pit and worked the Rhondda No.3 seam, which gained a world-wide reputation for the finest coking coal. In 1848, thirty-six coking ovens were built on the site, at a cost of £1,440. Demand for this coal was such that, in 1851, the Upper Cymmer Colliery was sunk and in 1855 the New Cymmer Colliery was opened alongside the Cymmer Old Pit.

However, in 1856 a major explosion occurred at the colliery, claiming the lives of 114 of the 119 miners working in the district of the explosion. It is sad to note, that thirty-eight young men under the age of sixteen were killed in this disaster. At the official enquiry, the cause of the explosion was found to be a build up of gas in the west side of the mine workings that caused a massive explosion due to inadequate ventilation throughout the mine. The Jury returned verdicts of manslaughter against the Officials of the colliery i.e. Jabez Thomas, brothers Rowland Rowlands and Morgan Rowlands, William Thomas and David Jones. The trial took place on 7 March 1857, at the Glamorgan Spring Assizes, under Judge Sir William Watson. The case against William Thomas and David Jones was dismissed and, although the others were put on trial, the Judge appeared to act in their interest. He directed the jury to discharge Jabez Thomas and advised them to quit the Rowlands brothers, which they did. The people of the community and especially the families of those killed felt angry, frustrated and bitter at the verdict.

After the trial the Cymmer widows and families had to struggle on without help, apart from the final humiliation of seeking parish relief and the receipt of the small sums doled out by the Cymer Widows and Orphans Fund, a public charity. Little wonder that such bitter feeling prevailed in the community for many years after.

In 1862 Insole made an unsuccessful attempt to reach the deeper seams of steam coal. However by 1877, in conjunction with his son, he succeeded in deepening the pit and opened four new seams of steam coal, the deepest being the nine foot seam, at a depth of 370 yards. He replaced the wooden headgear with an iron lattice structure and added a new upcast shaft to improve ventilation, in order to work the new steam coal seams in safety. In 1913, the colliery employed over 2,000 miners and produced over 11,000 tons of coal per week.

Due to a reduction in demand for this type of coal, mainly as a result of the introduction of oil fired vessels in the Royal and Merchant Navy, the colliery finally closed in 1940. The site was later cleared and in 1987 Leo's Supermarket & car park (later renamed Pioneer Supermarket) were constructed on the reclaimed site. In 1988, a memorial dram, mounted on rails, was dedicated by the Right Hon. Viscount Tonypandy PC, DCL, to the memory of all the miners who lost their lives as a result of their employment in the Cymmer Collieries.

A typical advertisement issued by a local colliery.

INSOLES LIMITED,

PROPRIETORS OF

CYMMER MERTHYR

Smokeless Steam Coal,

Supplied to the Principal Steamship Lines including—

The Peninsular & Oriental Steam Navigation Co.,
Compagnie Generale Transatlantique,
White Star Line (Ismay, Imrie & Co.),
Messageries Maritimes de France,
Austrian Lloyds Steam Navigation Co., etc., etc.,
And to all the principal Coaling Depots in the Mediterranean, East, South America, Brazils, etc.

ON THE ADMIRALTY LIST FOR USE IN THE BRITISH NAVY.

INSOLES NO. 2 RHONDDA COAL

Particularly Recommended for Bunkers.

Sole Shipping Agents :—

GEO. INSOLE & SON,

BUTE DOCKS, CARDIFF.

London Office—*23, LIME STREET.*

Telegraphic Addresses—"INSOLES, CARDIFF."
"INSOLES, (GRACE), LONDON."

A view of the Cymmer Colliery, Porth, *c.*1908.

A close-up view of the Cymmer Colliery.

A view of the Pioneer supermarket situated on the site of the Cymmer Colliery.

Cymmer Colliery, Porth, 15 July 1856.

On the morning of 15 July 1856 the most fearful and destructive explosion ever to have taken place before occurred at the Old Pit, Cymmer (Porth), situated at the gateway to the Rhondda Valleys, resulting in the loss of human life never experienced before in the history of coal mining, not only in Wales, but in the whole of Britain.

At 6 a.m. on that fateful Tuesday morning in July, 160 men and boys descended the shaft. They were on their way to their respective working places when a terrific explosion took place. At first it was thought that the everyone who had descended into the pit had lost their lives, but it was soon discovered that several miners had only penetrated a short way into the headings and were able to return to the pit bottom. By Tuesday evening 112 bodies, wrapped either in blankets or canvas, had been brought up from the pit. One body was recovered on the following day and one of those who had been rescued, but was severely burnt, died on the Thursday. In all 114 men and boys had died in this terrible tragedy.

The names of those who died are:

David Andrews, 26 years
William Collocott, 12 years
William Davis, 12 years
Daniel Daniel, 25 years
Jenkin David, 32 years
Morgan David, 18 years
David Davies, 15 years
Thomas Davies, 17 years
Thomas Davies, 48 years
William Davies, 28 years
Edward Davis, 10 years
Eli Davis, 10 years
Hezekiah Davis, 37 years
Thomas Davis, 10 years
Thomas Davis, 11 years
William Edmunds, 11 years
Samuel Edwards, 39 years
Thomas Edwards, 12 years
Benjamin Evans, 14 years
John Evans, 12 years
John Evans, 16 years
John Evans, 18 years
Mathew Evans, 60 years
Morgan Evans, 15 years
Morgan Evans, 16 years
Nathaniel Evans, 57 years
Philip Evans, 23 years
William Evans, 17 years
William Evans, 46 years
Benjamin Gibbons, age unknown
George Griffiths, 18 years
Henry Griffiths, 13 years
Peter Griffiths, 15 years
David Harris, 15 years
William Hayes, 12 years
Thomas Hayman, boy (age unknown)
Thomas Hopkin, 26 years
David Howells, 17 years
Edmund Howells, 22 years
Evan Hugh, 24 years
John Hugh, 18 years
William Hugh, 14 years
John Isaac, 17 years
R. James, 15 years
Henry Jarvis, 40 years
David Jenkins, 10 years
Rees Jenkins, 18 years
William Jenkins, 12 years
John Jenkins, 18 years
David John, 37 years
Isaac John, 18 years
J. John, 11 years
Joseph John, 64 years
Lazarus John, 10 years
Enoch Jones, 22 years
Henry Jones, 18 years
Thomas Jones, 10 years
William Jones, 11 years
Daniel Lewis, 30 years
David Lewis, 22 years
Edmund Lewis, 39 years
Evan Lewis, 26 years
Thomas Lewis, 12 years
Thomas Lewis, 30 years
William Lewis, 33 years
William Lewis, 36 years
Thomas Llewellyn, 28 years
William Martin, 23 years
Thomas Mathews, 40 years
Mathew Miles, 14 years
Richard Miles, 10 years
Walter Miles, 17 years

David Morgan, 14 years
John Morgan, 15 years
Isaac Morgan, 18 years
Morgan Morgan, 42 years
Morgan Morgan, 18 years
William Morgan, 11 years
Thomas Morgan, 12 years
David Morgan, 23 years
David Morgan, 17 years
Evan Phillips, 45 years
David Powell, 13 years
Aaron Rees, 23 years
Benjamin Rees, 42 years
John Rees, 11 years
Richard Rees, 27 years
Thomas Rees, 36 years
William Rees, 41 years
David Richards, 34 years
Hezekiah Richards, 24 years
John Roberts, 11 years
? Rudway, age unknown
Jacomi Salathiel, 18 years
John Salathiel, 14 years
George Soloway, 66 years
Henry Soloway, 13 years
Daniel Thomas, 21 years
David Thomas, 29 years
Gomer Thomas, 11 years
John Thomas, 18 years
Llewellyn Thomas, 13 years
Rees Thomas, 16 years
William Thomas, 19 years
Daniel Williams, 17 years
John Williams, 48 years
Thomas Williams, 38 years
Thomas Williams, 15 years
Thomas Williams, 34 years
William Williams, 28 years

Also:
The two sons of Hezekiah Davies, and the son of Joseph John (old John).

Left and below: The memorial to the miners who lost their lives in the explosion of 1856 was dedicated to their memory and officially inaugurated by the Rt. Hon. Viscount Tonypandy in November 1988.

Tynewydd Colliery, Porth

This colliery was sunk by James Thomas, Cope and Lewis of the Troedyrhiw Coal Company in 1852, to the No.2 and 3 Rhondda seams, which were reached at a depth of 270ft. In April 1877 an inrush of water from the Cymmer Colliery (Hindes Pit), which was situated on the other side of the river, flooded the workings and fourteen miners were trapped by the water. Five of the miners lost their lives, and four others were rescued eighteen hours later, leaving the remaining five miners trapped in a heading.

A party of volunteers immediately started rescue operations by cutting a small tunnel through the coal barrier, in a downward direction, to reach the trapped men. By this time the rescuers were aware of the problemscaused by the high pressure of the air in the vicinity of the trapped men. It was through the experience and ingenuity of Mr T.E. Wales, Inspector of Mines, that, on the tenth day, the rescuers broke through to the men who were safe, although they were suffering from the bends due to the rapid decompression which had taken place during the rescue attempt.

A local newspaper correspondent telegraphed the news to London, so that the whole of the country became aware of the plight of these miners and the efforts made to rescue them. Queen Victoria sent telegrams requesting to be kept informed of the rescue operations, and questions were asked in the House of Commons on the up to date situation.

Due to the bravery displayed by the rescuers, the constitution of the Albert Medal, which had so far only been awarded for bravery in the saving of life at sea, was altered so that it was considered for similar acts of bravery on land. Consequently, four of the rescuers were awarded Albert Medals, First Class and nineteen others were awarded Albert Medals, Second Class. (The story of this disaster is recorded in the book entitled *Disaster at Tynewydd* by Ken Llewellyn.)

The colliery closed in 1901 and the site was dismantled and cleared in 1906. In 1907 a large hotel, appropriately called the Tynewydd Hotel, was built on the site, together with a range of other commercial buildings. In the 1920s Dan Jones & Son, a fruit and vegetable wholesaler occupied another part of the site. The site was taken over by a motor repair company for many years but closed in 1998. It was then taken over by a new skateboard and childrens Playpark with some of the older industrial sheds abandoned. The Tynewydd Hotel was converted to a night-club in 1999.

In 1860 financial problems led George Insole to lease the Upper Cymmer Colliery to T.C. Hindes, of Neath. The miners working at this colliery lived in the small cottages situated only a few yards from the pithead. In 1877 water from the this colliery flooded the nearby Tynewydd Colliery.

A view of Tynewydd Colliery in the 1880s.

The site of the Tynewydd Colliery today.

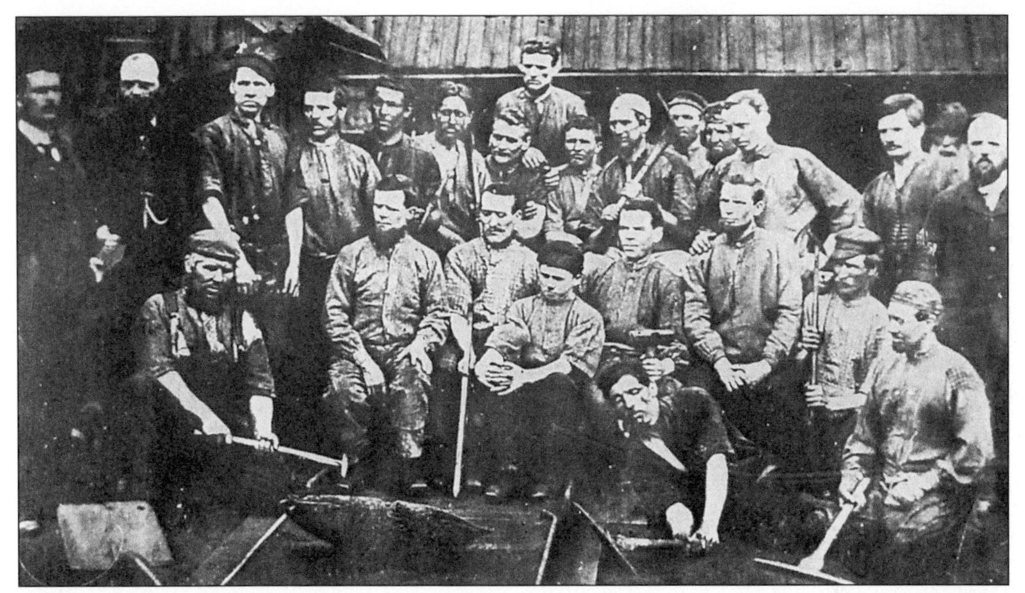

The rescued and rescuers at the Tynewydd Colliery. The boy David Jones is seated in the middle of the rescued miners. The other rescued are Moses Powell, George Jenkins, John Thomas and David Jenkins. Standing at the extreme left are Daniel Thomas and his brother Edmund Thomas who later received the medals.

Left: The Clanny Lamp used by Mr Edmund Thomas in the rescue of the Tynewydd Miners.

Right: The Royal Albert Medal, First Class was awarded to Mr Daniel Thomas, colliery proprietor, Brithweunydd, for bravery shown during the Tynewydd Disaster at Porth. This medal was presented to him in the presence of 40,000 people, on the Pontypridd Common, by Lord Aberdare, on behalf of Queen Victoria, on 4 August 1877. The Royal Albert Medal, Second Class was awarded to Mr Edmund Thomas, colliery owner, Llwyncelyn. Both medals read 'For gallantry in saving life on land.'
(Photographs by courtesy of Mr Cecil Thomas, Cardiff).

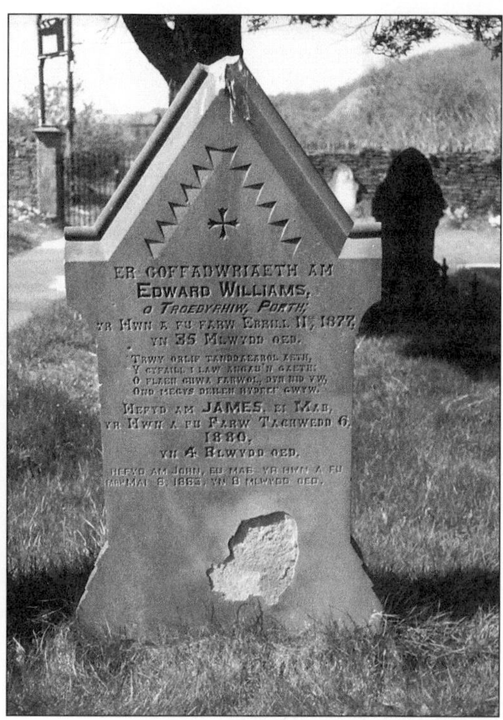

Above and left: Graves of some of the victims of the Tynewydd disaster: Robert Rogers, aged thirteen; Edward William, aged thirty-five; John Hughes, aged forty-five and his son William Jones Hughes, aged eighteen.

Five

Rhondda Fawr

The shop of the Jones' Brothers butchers, at Clydach Vale, was damaged in a riots during the miners strike of 1910.

Dinas (A Fort or City)

Dinas Colliery, Dinas.

In 1809 Walter Coffin bought Dinas Uchaf farm and gave notice to the tenant Lewis Robert Richard to terminate his tenancy so that he could prospect for coal on his estate. His first level led him to the No.1 Rhondda seam but the coal was extremely difficult to access and of such an inferior quality that Coffin opened another level leading to the No.2 Rhondda seam. The coal, in this instance, was of a good quality and easy to extract from the level, which prompted him to expand his enterprise.

In 1810 and 1811 he leased further mineral rights in the area and, in 1812, sunk the first deep mine in the Rhondda, at the Dinas Lower Colliery, to the No.3 Rhondda seam, which was found at a depth of 40 yards. This seam of coal was of the highest quality, being in fact one of the finest coking coals in South Wales at that time. Although marketed as Dynas No.3, this soon became known as Coffins coal.

In 1832 the Dinas Middle Colliery was sunk 600 yards higher up the valley and the No.3 Rhondda seam was found at a depth of 80 yards. Output continued to flourish and, by 1841, 430 men and boys were employed at Dinas.

On 1 January 1844 an explosion took place causing the deaths of twelve men and boys. This was the first major explosion to take place in the Rhondda Valleys and was not, unfortunately, going to be the last. At the inquest the coroner severely censured the overman Griffith Williams for taking the young Edmund Llewellyn with him and allowing him to test for gas on that fateful day, which later resulted in the explosion. The coroner invited the jury to record a verdict of manslaughter but the jury declined to do so, as the young lad was the son of the manager, Daniel Llewellyn.

In 1869 a new shaft was opened by the owner, Daniel Thomas, in order to improve conditions and to increase production. Unfortunately, on the 13 January 1879, a terrific explosion took place, resulting in the deaths of sixty-three men and boys. The official report on the cause of the explosion stated that the origin of that explosion was accidental and that it occurred through a sudden outburst of gas in or near Morriss heading.

In 1881 the new owners, the Dinas Steam Coal Company, deepened the shaft to reach the lower steam coal seams. Production of this coal remained fairly constant until 1887, then rapidly declined to a low output, until the colliery closed in 1893.

Dinas Colliery, Dinas, 1 January 1844.

On 1 January 1844, the day shift had gone down the pit and at approximately 8.00 am. the first major explosion in the Rhondda Valleys took place, causing the deaths of twelve men and boys. At the moment the explosion took place, the manager Mr Daniel Thomas, together with a young lad, David Morgan were descending the shaft. The pithead gear was destroyed by the explosion and the cage plunged to the pit bottom. Daniel Thomas was injured, but recovered, David Morgan was not so fortunate and died from his injuries.

Thomas Rowlands who, together with his son David, was killed in the disaster, was the brother of Moses Rowlands, schoolmaster of Penderyn, who came to Dinas at the invitation of Walter Coffin. He was later appointed Clerk of Works at Coffins' Collieries. These are the names of those who died:

William Harrhy, 20 years
David Job, 14 years
Thomas Leyshon, 16 years
David Morgan, 14 years
Lewis Morriss, 12 years
Thomas Morriss, 48 years
William Morriss, 17 years
David Phillips, 22 years
Edward Powell, 21 years
John Richards, 14 years
Thomas Rowlands, 61 years
David Rowlands, 9 years

Dinas Colliery, Dinas, 13 January 1879.

On 13 January 1879, just before midnight, a terrific explosion took place at the colliery shaking not only the colliery buildings but also the nearby houses. Dense clouds of smoke and multi-coloured fumes billowed from the main shaft. At this time the colliery consisted of two pits, the lower pit (the downcast) and the middle pit (the upcast). Rescuers descended through the lower pit, heading towards the middle pit but, due to the extensive damage caused by the explosion and the presence of gas, operations were suspended until the ventilation at the colliery could be improved. Sixty-three men and boys were killed in this explosion. Many were buried in a mass grave at Llethr-Ddu Cemetery at Trealaw.

It is interesting to note that even in such troubled times, Daniel Thomas, owner of the Dinas Colliery, had to approach the Joint Burial Board for the parish of Ystradyfodwg for permission to intern the bodies recovered from the explosion on a Sunday. Needless to say permission was granted. The names of those who died are listed below. Their ages are unknown.

James Bowen
Robert Chubb
William Cross
Evan Davies
Edward Davies
Daniel Dunn
Daniel Dunworth
James Edwards
John Edwards
Robert Emery
Joseph Evans
William Evans
John Griffiths
John Griffiths
William Griffiths
William Griffiths
James Harris
Jebu Hawkins
Henry Hayter
Thomas Holmes
Richard Howells
David Jenkins
Evan Jenkins
John Jenkins
William Jenkins
Evan John
Elijah John
William Jones
River Jordan
John Landrigan
John Lewis
William Lloyd
Isaac Martin
Charles Meade
Frank Moore
Daniel Morby
Charles Penny
Samuel Prior
David Rees
Edward Rees
Thomas Rees
Thomas Richards
William Richards
Thomas Roberts
William Roberts
John Rounsevell
Samuel Rounsevell
Daniel Smith
Edward Sullivan
Henry Taylor
David Thomas
Charles Westlake
Charles Wheadon
Octavius Wheadon
Henry Williams
Henry O. Williams
Lewis Williams
Stephen Williams
William Williams
William Williams

In Sad Remembrance of

63 MEN AND BOYS

WHO WERE

KILLED IN THE MIDDLE PIT,

DINAS COLLIERY, RHONDDA VALLEY,

BY AN EXPLOSION,

On Monday Night, January 13th, 1879.

Of this number there were 46 married and 11 single men, 3 boys under 18, and 3 widowers. The dependent relations are:—46 widows, 130 Children; 2 parents, and 3 sisters.

"DEATH COMES IN ALL SHAPES."

Copy of the memorial card issued in memory of the victims of the Dinas Colliery disaster.

The mass grave at Llethr-Ddu Cemetery, Trealaw.

The scene at Dinas Colliery after the explosion in 1879.

Brithweunydd Workingmen's Home, Dinas, commonly known as the Lodging House. It was built by Walter Coffin (c.1832) for the sinkers of the Dinas Middle Coliery and was later used by the single miners employed at the colliery.

The childrens playground on the site of Walter Coffins' Dinas Lower Colliery.

Dinas Middle Colliery, 1879. The Dinas and Tonypandy railway station is visible in the foreground.

The site of the Dinas Middle Colliery today. The capped shaft is visible in front of the third block of flats.

Penygraig (End or Top of the Rocks)

Ely Pit, Penygraig

Sunk in 1892 by the New Naval Colliery Company which became part of the Cambrian Combine in 1908. In 1910 it employed 939 men and was the scene of the start of a dispute which would have serious consequences throughout the South Wales Coalfield for many years. The trouble began over the management offer of a piece rate for the new Bute seam which had been recently opened. The miners demanded more money because this particular seam was difficult to work.

On 1 September 1910 the owners locked out not only the miners in dispute but all the workmen employed at the colliery, a total of 950 men. On 1 November after a ballot, the South Wales Miners' Federation called out the whole 12,000 miners employed by the Cambrian Combine. Great bitterness developed in the Mid-Rhondda area, especially when it became known that the owners were to employ black-leg labour to keep the pumps and ventilation going at the Glamorgan Colliery, Llwynypia. This resulted in skirmishes and riots between police and miners. The strike continued until the autumn of 1911 when hardship and misery forced the miners to accept the coal owners' offer.

Although activities restarted in the collieries, there were not enough places available to house some 3,000 miners. By 1912, the numbers of miners employed at this pit had been reduced to 650. The colliery continued to produce coal until around 1928, when activities ceased and the pit was only retained for pumping purposes. After nationalization in 1947, it continued to be used for pumping purposes until final closure in 1958, together with the other pits comprising the Naval Complex.

In 1987, during the construction of the new Tonypandy bypass, part of the old colliery site was used in the construction of the new road. This was not only beneficial to the people of the Rhondda but gave better access to the new Ely Industrial Estate. The bypass, which cost £6.7million, was officially opened by the Rt. Hon. Peter Walker, Secretary of State for Wales, on 5 October 1988. Further developments included the laying of a new football pitch to replace the old one, together with the general landscaping improvements to the area.

The Ely Colliery, 1920.

The site of the colliery today.

Tonypandy (Lay/Land of Fulling Mill)

The Naval Colliery.

The Rowland family of Penygraig had been involved with mining in the area for a considerable time, first with the opening of a level at Penygraig in 1857 and then with a small colliery working, the No.3 Rhondda seam, in 1864. Around 1875 Moses Rowlands Jnr, the chief partner in a company known as the Naval Colliery Company, opened the Pandy Pit, the owners being Morgan and Rowlands. Its main coal seams were reached in 1879 and, in the next year, over 76,000 tons of good quality steam coal were raised from this colliery.

Tragically, in 1880 a serious explosion occurred and 101 miners lost their lives. At the inquest the official verdict was that this explosion took place on or near the Turberville heading and was apparently caused by a fall nearby. This released a quantity of gas which, mixed with air, led to an explosion causing the mixture, at high velocity, to come in contact with a lamp and ignite by blowing the flame through the gauze. The inquest finally declared that the cause was accidental.

In 1884 another explosion took place and eleven miners and three of the rescue party lost their lives. It is interesting to note that Mr Daniel Thomas, the proprietor of the Dinas Colliery and holder of the Albert Medal, First Class, whilst attempting the rescue of the eleven men trapped in the pit, succumbed to after damp and died.

The official verdict at this inquest was as follows: 'We agree that the deaths of the eleven miners was accidental. It was caused by an explosion. We are also agreed that the explosion was caused by the shot fired by the late William Williams but we beg to add that the Manager should have given the shotman more explicit orders in firing the shots at that particular time and place. The verdict on the rescuers was given as Death by suffocation resulting from the explosion.'

After these explosions the Company was sold in 1887. The new owners, W.O. Hurt (Dinas) and Messrs Wood and Gregson (Cardiff), renamed it the New Naval Collieries Company and extended the area under the mineral lease, opening three new collieries, the Ely (1892), Nantgwyn (1892), and the Anthony (1910) pits. This development increased the number of men employed from 800 in 1890 to 2,640 in 1908, when the company became part of the Cambrian Combine. During the same period the output tonnage increased from 128,000 to 595,000 tons per annum.

Due to the decline in the demand for coal in later years, the pit closed in 1958. The power house and the 80ft stack were demolished in December 1962. The site was then partially cleared and certain remaining colliery buildings were converted into small industrial units for use by local industry. The large feeder unit was filled in and became the site of a garage and petrol station. The remainder of the site was developed in 1967, with the construction of a rugby ground for the Penygraig Rugby Football Club, which included the erection of a covered stand, floodlighting and raised terraces for approximately 5,000 spectators. The ground was officially opened in December 1970.

In 1973, in conjunction with the Nantgwyn Project, the massive coal tip was lowered by 150ft and the tip spoil, estimated to be 1.6 million tons, was removed and transported for use at the Nantgwyn site, which had been approved as the site for the new Tonypandy Comprehensive School.

The Judges Hall, Tonypandy, built in memory of Judge Gwilym Williams of Miskin Manor, was officially opened by HRH the Princess Louise accompanied by the Duke of Argyll on Friday 23 July 1909. After this opening the Princess, accompanied this time by Mr L.W. Llewellyn, General Manager of the Cambrian Combine, officially opened the aerial incline which transported the colliery waste to the top of the mountain. She stated that she hoped it would be successful and beneficial to the people of the Rhondda. Her Royal Highness and her party then left, amid many cheers from the assembled crowd.

A panoramic view of the Naval Colliery site today.

Naval Colliery, *c*.1920.

Entrance to the Naval Colliery. The lamp room is in the middle-right.

Naval Colliery, Penygraig, 10 December 1880.

At 1 a.m. on Friday 10 December 1880, the first indication that a disaster was about to take place was a low rumbling in the colliery site. Suddenly, a blast burst out of the main shaft of the mine, spilling sulphur fumes and smoke into the surrounding area. The rescuers experienced difficulties and it took them twelve days to bring all those who perished in this terrible disaster to the surface. Only then was it fully realised that 101 miners had died in this explosion.

The names of those who died:

Henry Brooks, 23 years
James Chadwick, 25 years
Archibald Cook, 27 years
Evan David, 49 years
Evan David, 32 years
John Davies, 42 years
John Davies, 25 years
William Davies, 26 years
Thomas Edwards, 59 years
W.R. David, 23 years
David Evans, 26 years
David Evans, 23 years
Howell Evans, 45 years
John Evans, 35 years
William Evans, 39 years
James Francis, 50 years
Alfred Fry, 30 years
William Fry, 27 years
Zephaniah Gibbon, 36 years
Griffith George, 24 years
James Gibbon, 20 years
Thomas Grice, 21 years
William Grice, 23 years
Evan Griffith, 22 years
William Howell, 21 years
Benjamin Howells, 52 years
Edward Hughes, 42 years
John Hughes, 29 years
Henry Isaacs, 34 years
Thomas Isaacs, 28 years
John Jenkins, 40 years
John Jenkins, 54 years
John Jenkins, 54 years
Evan Jones, 45 years
Rees John, 19 years
Henry Jones, 32 years
John Jones, 35 years
Joseph Jones, 30 years
Thomas Jones, 21 years
Thomas Jones, 31 years

Thomas Jones, 36 years
Thomas H. Jones, 25 years
David Lewis, 43 years
Edward Lewis, 42 years
James Lewis, 15 years
Rd. Lewis, 24 years
Samuel Lewis, 14 years
William Lewis, 56 years
William Lewis, 49 years
David Lloyd, 70 years
John Lloyd, 40 years
David Lodwick, 45 years
William Maskman, 26 years
John McCarthy, 32 years
Edward Morgan, 66 years
James Morgan, 75 years
Joseph Morgan, 25 years
Thomas Morgan, 17years
Thomas Morgan, 36 years
Joseph Morris, 32 years
James Murphey, 50 years
William Parlour, 29 years
William Pearce, 37 years
Evan Phillips, 19 years
Eli Raps, 29 years
Evan Ratcliffe, 30 years
Thomas Reed, 29 years
Evan Rees, 27 years
John Rees, 34 years
David Roberts, 31 years
Robert Roberts, 16 years
William Roblin, 40 years
Noah Rodgers, 39 years
David Rosser, 20 years
George Samuel, 30 years
Phillip Samuel, 18 years
Samuel Samuel, 14 years
Thomas Simon, 23 years
John Snook, 52 years
John Stone, 48 years
David Thomas, 33 years
John Thomas, 18 years
Morgan Thomas, 25 years
Thomas Thomas, 31 years
David Watkins, 44 years
David D. Williams, 36 years
David Williams, 59 years
Evan Williams, 35 years
Griffith Williams, 22 years
James Williams, 38 years
John R. Williams, 23 years
Richard Williams, 49 years
Thomas Williams, 40 years

Naval Colliery Tonypandy, 27 January 1884.
On Sunday morning, eleven officials of the colliery had descended into the mine to prepare for the restart of production the following day. Suddenly, there was a terrific explosion which resulted in the death of not only these officials, but some of the members of the rescue team.

The names of the miners who died are listed below. Their ages are unknown.

Thomas Davies
Solomon Edwards
John Heycock, father
John Heycock, son
Oliver John
David Jones
Fred Nugent
John Price
John Rees
James Seville
William Williams.

Those of the rescue team who died:

Edward Watkins, of the Adare Inn, (known as Ned Adare),
Thomas Lewis, under-manager,
Daniel Thomas, owner of the Dinas Colliery and holder of the Albert Medal First Class.

Daniel Thomas. Awarded the Albert Medal, First Class, for bravery at the Tynewydd Colliery Porth in April 1877, he died while attempting to rescue miners after the explosion at the naval Colliery Penygraig on the 27 January 1884.

A view of the Penygraig Rugby Club situated on the site of the Naval Colliery. The first floodlit match played on this ground was on 8 October 1970 between Penygraig and Pontypridd.

A view of the Anthony pit, sunk in 1908 on the site of the Naval Colliery and merged with the existing Pandy pit in 1913. It closed in 1952.

A view of the site of the Anthony pit today.

Nantgwyn Colliery, Tonypandy.

Nantgwyn Colliery was originally sunk in 1892 by the Naval Colliery Company, which in 1908 became part of the Cambrian Combine. By 1910 Nantgwyn employed 821 men, a figure which had increased to 1,150 men by 1913. It continued in production until around 1928, when it ceased to produce coal but was retained for pumping and ventilation purposes. It was taken over by the National Coal Board in 1947 and continued in this capacity. In 1952 Nantgwyn was merged with the Naval Colliery but final closure came in 1958.

In 1973, a massive land reclamation scheme for the Nantgwyn and the Naval Colliery sites was commissioned; once completed it would provide the site for the new Mid-Rhondda Comprehensive School. At a cost of £400,000, the scheme, to be completed in twelve months, would cover an area of 100 acres of tip and surrounding hillsides. After the necessary preparations had been carried out, the upper Naval Colliery tip was to be lowered in height by 150ft and 1,600,000 tons of tip material would be transported to various parts of the Nantgwyn site to provide infill to secure 20 acres of flat land for playing fields for the new school.

The remainder of the site was then landscaped and grassed over to blend in with the surrounding mountains. In 1978, the new Comprehensive School was built on this site and the project was completed in January 1979.

A view of the Nantgwyn Colliery, 1925.

Another view of the Nantgwyn Colliery from the 1930s.

The colliery site today. This view shows the comprehensive school and the landscaped area.

Clydach Vale (Vale of the rocky bed)
Blaenclydach Coliieries, Clydach Vale.

The Blaenclydach levels were opened by Frank James, in 1863, to work the Rhondda No.3 seam. Good quality coking coal continued to be extracted until 1875. During this year the Blaenclydach Colliery was opened, the owners at that time being Mr Bush and Company. However, in 1878, the colliery was purchased by the Forest Iron and Steel Company, in order to ensure that a continuous supply of good quality coking coal was available to feed their Pontypridd furnaces, engaged in the production of pig iron. Around 1912 the Blaenclydach Drift Mine, situated near the old abandoned Cwm Clydach Colliery which had closed in around 1895, was opened by the Blaenclydach Colliery Company.

In 1913, records show that 726 men were employed at this new Blaenclydach Colliery, known locally as the Gorki Drift Mine, while only twenty men remained employed at the Old Blaenclydach Colliery, whose owners at that time, the Brookvale House Coal Company, changed its name to Brookvale. In 1920, the whole complex was bought by the Rhondda Coal Company who, at that time, employed 385 men. In 1940 the complex was purchased by the Powell Duffryn Company, who remained its owners up to the time of nationalization, in 1947.

On 25 November 1941 a tragic accident occurred at the colliery, resulting in the loss of seven lives, while fifty-three men were injured. It would appear that about ninety men were being conveyed into the drift on a journey of flat bottom drams fitted with seats (commonly known as a spake) when it ran wild for a distance of about 200 yards before it was stopped. The impact caused the resulting loss of life and injury.

Shortly after nationalization in 1947 the colliery was closed by the National Coal Board, with a loss of forty-one jobs. Many years later, in 1985, the derelict site was incorporated into a massive landscaping project in conjunction with the Cambrian Colliery and carried out by the Cambrian Colliery Reclamation Scheme, at a cost of £2.5m. The area was contoured and landscaped into the attractive leisure area as it is today.

In April 1994, a memorial stone marking the disaster of 1941 was unveiled on the site by George Rees, Area General Secretary of the NUM. Since Local Government reorganization in 1996, part of the site has been used to house the main administrative headquarters of the newly formed Rhondda Cynon Taff County Borough Council.

The reclaimed site of the Blaenclydach Colliery today. The covered over-shaft is still visible near the car park.

The hollow in the mountainside was the site of the Cwm-Clydach Colliery, opened in 1864 by Daniel Thomas Sr and his son Edmund. The colliery closed in 1895 and remained abandoned for many years. The whole area was eventually landscaped to provide small factory units and sporting facilities for the area. It was finished off by the planting of trees by the Forestry Commission.

Blaenclydach Coliery, c.1920. The drift-mine was opened around 1912 and was locally known as the Gorky Drift Mine. The Bush houses are visible on the right and, in the background, the headgear of the abandoned Cwm-Clydach Colliery.

Blaenclydach Colliery, Blaenclydach, 25 November 1941. (Gorki Drift Mine)
On the 25 November 1941, at 7a.m., a group of miners clambered aboard flat bottom trams, known as spakes. Fitted with seats the spakes were designed to descend into the mine workings. About 200 yards into the mine the spake careered out of control, creating panic among the miners on board. Whilst some miners leapt for safety, others prepared themselves for the inevitable collison. When this occurred, six men were killed, while fifty-three were badly injured.

The names of those who died are:

Kenneth Francis, 19 years
William Gregory, 16 years
Len Hatcher, 15 years
Dennis Humphries, 19 years
Elias Price, 18 years
Robert Williams, 43 years

Ken Owen, one of those badly injured, later died from his injuries. This brings the total of those who died in the disaster up to seven.

The Memorial Stone on the site of the disaster. The administration buildings are visible in the background.

Cambrian Colliery Clydach Vale.

In 1871 Samuel Thomas and J. Osborne Riches created the Cwmclydach Colliery Company to work the steam coal seams at the top of the Clydach Valley. With the help of Thomas Joseph (the brother in law of Samuel Thomas) from Aberdare, they sunk the No.1 shaft in 1872 and, in 1874, they struck the 6ft seam at a depth of 400 yards. This was regarded as one of the finest quality seams of steam coal in the South Wales coalfield at that time.

The new No.2 shaft was opened in 1875 and, when Samuel Thomas died in 1879, the two collieries were producing a combined total of over 1,000 tons of coal a day. In order to transport this coal to the docks, the company constructed, in 1884, their own private railway down the steep valley incline to the Pwll yr Hebog Junction, to link up with the existing Taff Vale Railway on the valley floor. This link was bought by the TVR in 1890.

In 1889 the No.3 shaft was sunk. Completed in 1891, by the end of the century over 3,000 tons of coal were produced daily from these three pits. As the night shift descended on Friday 9 March 1905, a loud explosion was heard from a quarter of a mile below ground. The ensuing disaster killed a total of thirty-three miners. At the inquest, and from the evidence which had been submitted, the opinion was that the explosion occurred in the No.6 cross cut of the No.5 District, in consequence of something happening to David Enoch's lamp. Therefore no blame can be attached to anyone for the explosion.

After nationalization in 1947, the pits continued to flourish but, in 1965, like many other collieries in the coalfield, Cambrian was coming to a close. Half of the workforce of 800 miners had either finished or had transferred to other higher output collieries. At 1 p.m. on Monday 17 May 1965, a terrible explosion at the colliery, resulted in the deaths of thirty-one miners. The inquest found that the explosion was caused by a ventilation problem which allowed pockets of gas to accumulate. This was ignited by an electric arc at an open switch panel on which electricians were working to locate a fault, which had brought the face machinery to a halt. This was the last major mining disaster in Wales.

The grief caused by the scale of this disaster, and the resultant devastation for the local community, was not only felt throughout the mining valleys but in the entire country. A relief fund was set up with donations coming not only from this country but also from Europe. The colliery closed in 1966. In 1985 the Cambrian Colliery Reclamation Scheme started, with the demolishing of the last remaining buildings on the site by means of explosives.

This signalled the beginning of a £2.5 million Land Reclamation Scheme. This scheme was one of the biggest colliery site reclamations attempted and covered an area of 280 acres. Work involved the movement of 2 million tonnes of colliery waste spoil. As no fresh soil was brought in, the existing soil needed to be reshaped. In 1992, a memorial service took place. The site, incorporating a pit head wheel and memorial garden, was dedicated to the memory of miners who had died in both disasters and was officially opened by Neil Kinnock MP, in the presence of many of the widows and families of those who had died, together with local dignitaries.

Pwllyrhebog Incline

The Pwllyrhebog Incline, opened in 1863, has a gradient of 1:13. It was originally operated by a balance rope and pulley system, with horses being employed at the top and bottom of the incline. However, when locomotives came into being, those to be used at Pwllyrhebog were specially built and popularly referred to as Incline Engines.

They were of the 060 tank class engine, class H, and were numbered 141, 142, and 143. They were built in 1884 by a company called Kitson and worked continuously on the incline until it closed in July 1951, sixty-seven years later!

Pwllyrhebog Incline Engine No.141. (Photograph by courtesy of Mr H. Shadbolt, Trealaw.)

A view of the Cambrian Colliery, *c*.1904. Note the spelling of Clydach.

Another view of the Cambrian Colliery in 1905. This picture was taken after the explosion in which thirty-three miners died.

Cambrian Colliery, Clydach Vale, 10 March 1905.
A massive explosion took place on 10 March 1905 as the shift was changing at the colliery. Had the tragedy occurred a few minutes earlier there would have been about 550 men in the workings. As it was, thirty-three men were killed.

The names of those who died are as follows:

Joseph Chalker, 39 years
Frank Shallish, 45 years
Thomas Davies, 21 years
Jenkin Davies, 49 years
Evan Davies, 42 years
Noah Edmunds, 48 years
David Enoch, 35 years
William Evans, 45 years
Evan Evans, 36 years
John Griffiths, 40 years
William Griffiths, 60 years
William Gronow, 51 years
Morgan Harding, 47 years *
William Harris, 39 years
Henry Harvey, 45 years
Thomas Hawkins, 48 years
Thomas John, 36 years
Edward Jones, 47 years
Joseph Jones, 50 years
John Jones, 38 years
David Lewis, 55 years
Adam Lewis, 45 years
Rees Lewis, 53 years
William Lewis, 39 years
Thomas Morgan, 26 years
David Morgan, 35 years
John Owen, 50 years
Thomas Richards, 42 years
John Ridge, 52 years
Edwin Thomas, 33 years
William Tudball, 42 years
Owen Williams, 24 years
Robert Williams, 23 years.

*Morgan Harding was rescued alive, but died nine days later from burns.

The colliery site today, after the completion of the restoration work.

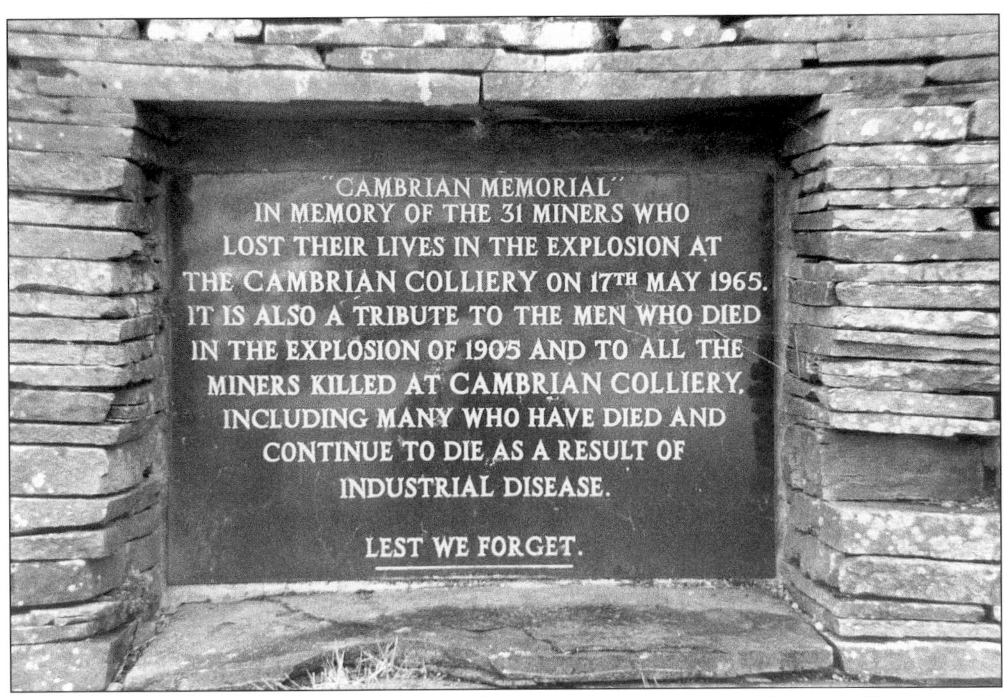

The Memorial to the miners who were killed while working at the Cambrian Colliery.

Cambrian Colliery, Clydach Vale, May 17 1965.

It was just after 1 p.m. on that spring day that the explosion occurred underground, claiming the lives of thirty-one miners. The colliery was earmarked for closure, and almost half of the total workforce of 800 miners had already been transferred to other collieries, otherwise there could have been more fatalities. Thankfully, this was to be the last explosion resulting in loss of lives in the Rhonnda Valleys, and the last major explosion disaster in Wales.

The names of those who died are:

Ronald Arnold, 48 years
Ernest Breeze, 38 years
Ernest Burnett, 46 years
Peter Calvert, 40 years
James Channing, 46 years
Albert Colcombe, 44 years
Raymond Daniels, 34 years
Gerrard Davies, 24 years
Kenneth Davies, 26 years
David Evans, 28 years
Ronald Flower, 45 years
Ronald Gregson, 28 years
David Griffiths, 43 years
Thomas Hann, 42 years
Richard Hucker, 32 years
Ivor Jacobs, 45 years
Henry Lee, 56 years
Leonard May, 33 years
Ivor Morgan, 32 years
Arthur Newman, 46 years
Vivian Nicholas, 51 years
Harold Pope, 50 years
Donald Price, 42 years
Evan Rees, 48 years
Richard Roberts, 55 years
Gwilyn Thomas, 28 years
William Thomas, 33 years
Edmund Williams, 51 years
Leslie Williams, 54 years
Sidney Williams, 47 years
Trevor Williams, 27 years.

The peace and tranquillity of an area which had previously witnessed the bedlam of noise and industrial upheaval and tragedy.

Llwynypia (The Magpie's Bush)

Llwynypia was known locally as Scotch Colliery, after the Scottish miners brought from Kilmarnock by Archibald Hood, when he sank shaft Nos 1, 2 and 3 in 1862.

Further developments took place with the sinking of additional shafts. No.4 was sunk as a pumping pit, No.5 was sunk as an upcast for No.3 shaft and No.6 was sunk and used as an upcast for Nos1 and 2 shafts. At this time the colliery was known as the Glamorgan Coal Company, but became part of the Cambrian Combine in 1908, under the Chairmanship of D.A. Thomas, Liberal M.P. for Merthyr, 1888-1910, who later became Viscount Rhondda.

The Cwrt Terrace, locally known as the Huts, was built for the sinkers of the colliery. Demolished in 1963, it was replaced with a new block of flats called Mitchell Court, named after Alderman Sidney Mitchell. One essential by-product of the colliery was the brickyard, producing bricks for the lining of shafts being sunk, various colliery buildings and the rows of homes known locally as the Scotch Terraces, which Hood had built to house his workers. In its heyday more than 10,000 bricks per day were produced, mostly by women using hand moulds. The bricks were then dried undercover on iron floor plates heated by colliery waste gasses by means of underfloor flues. The fireclay used for the bricks came from No.3 pit, while they worked the No.3 seam.

A massive explosion occured in 1932, causing the deaths of eleven men. Output at the colliery declined and coal extraction ceased in 1945, but it remained open for maintenance and pumping purposes until around 1966. About this time a pumping station was built, and the automatic pumps installed are still in use today.

The site was cleared of all buildings, with the exception of the Power House which became a listed building and still stands today. It was used as a base for massive police reinforcements during the Cambrian Combine Strike of 1910/1911. From outside the building striking miners were driven back by police to Tonypandy Square, where windows were broken and shops looted during the ensuing Tonypandy Riots. The area is the subject of modern development, and various small factories have been built on the site, together with a new fire station.

In 1994, the new Magistrates' Court was built on the site of the old colliery stables, situated at the side of the Glycornel Lake. In August 1995, Hutchings Motors transferred their business to the large empty factory at the colliery site. This enabled them to offer customers a better service at a more convenient location.

Glamorgan Scotch Collieries, c.1905.

Brick workers at Glamorgan Colliery, *c*.1880. The majority of workers were women.

Brick workers at the colliery in 1900. The demand for bricks was increasing, hence the need to modernise the plant. In its heyday 10,000 bricks were produced each day.

Glamorgan 'Scotch ' Colliery, late 1900s.

The colliery in its heyday.

A view of the Glamorgan Scotch Colliery, *c*.1904. Note the position of the kilns, used for the manufacture of the bricks, on the right.

The new Tonypandy by-pass which splits the site of the former Glamorgan Colliery. The flat area (top left) was the site of the Scotch No.6 pit.

The Power House, Llwynypia. The scene of many bitter struggles during the 1910 Cambrian Combine strike.

Glamorgan (Scotch) Colliery, Llwynypia, 5 January 1932.
An explosion at the colliery occurred at 7.30 p.m. on 5 January 1932, in the Pentre seam. Eleven miners, including two of the rescue party, were killed. It seems that the explosion, which took place in the Pentre seam of No.1 pit, was caused by a build up of gas, due to ventilation problems, and the cause of death in each case was carbon monoxide poisoning.

The names of those who died are listed below. Their ages are unknown.

Morgan Bowen, Miner
Charles Cryer, Fireman
Stanley Dando, Miner
Herbert Evans, Miner
David Hughes, Miner
David Rodgers, Miner
Cliff Sparrow, Miner
William Thomas, Miner

The rescue workers were John Evans, Miner, and John Jones, Fireman.

Both of these gallant men were posthumously awarded Certificates of Bravery by the Carnegie Hero Trust for Humanity.

For risking their lives while rescuing miners from the poisonous fumes which occurred after the explosion, the following miners were awarded the Albert Medal, Second Class:

Mr John Evans,
Mr Frank Cordey,
Mr ? Hughes.

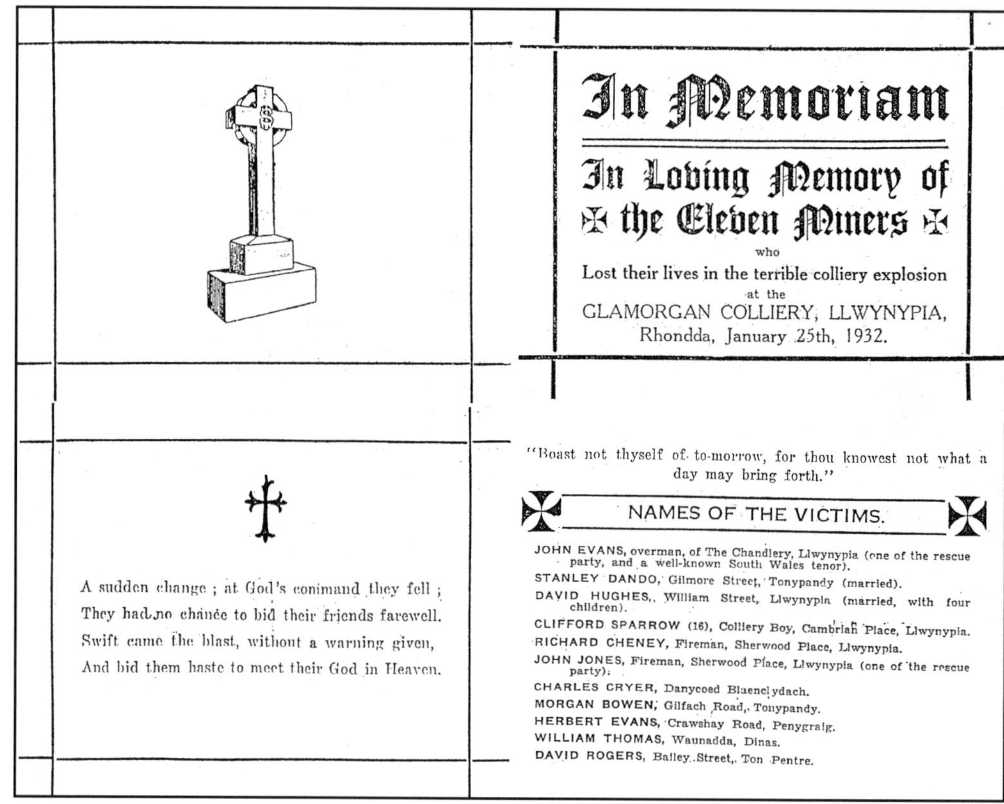

In Memoriam

In Loving Memory of
the Eleven Miners
who
Lost their lives in the terrible colliery explosion
at the
GLAMORGAN COLLIERY, LLWYNYPIA,
Rhondda, January 25th, 1932.

A sudden change ; at God's command they fell ;
They had no chance to bid their friends farewell.
Swift came the blast, without a warning given,
And bid them haste to meet their God in Heaven.

"Boast not thyself of to-morrow, for thou knowest not what a day may bring forth."

NAMES OF THE VICTIMS.

JOHN EVANS, overman, of The Chandlery, Llwynypia (one of the rescue party, and a well-known South Wales tenor).
STANLEY DANDO, Gilmore Street, Tonypandy (married).
DAVID HUGHES, William Street, Llwynypia (married, with four children).
CLIFFORD SPARROW (16), Colliery Boy, Cambrian Place, Llwynypia.
RICHARD CHENEY, Fireman, Sherwood Place, Llwynypia.
JOHN JONES, Fireman, Sherwood Place, Llwynypia (one of the rescue party).
CHARLES CRYER, Danycoed Blaenclydach.
MORGAN BOWEN, Gilfach Road, Tonypandy.
HERBERT EVANS, Crawshay Road, Penygraig.
WILLIAM THOMAS, Waunadda, Dinas.
DAVID ROGERS, Bailey Street, Ton Pentre.

Copy of the memorial card issued after the explosion.

Pentre (A Village)

Gelli Collieries

The colliery was sunk in 1877 by Edmund Thomas, a colliery owner of Porth, and George Griffiths, a grocer and draper of Pontypridd. Edmund Thomas built a large house known as Maindy Hall on the site of Maindy Farm, in Ton Pentre, so that he could see the collieries from his house.

In August 1883, an explosion took place at the colliery. One miner died instantly and twenty-one were severely burnt. Later, four of the miners died from their injuries, bringing the total of deaths to five. At the inquest it was stated that the explosion occurred due to a build up of gas, the cause of ignition was unknown. The verdict was accidental death.

Financial problems meant the collieries were sold, along with Tynybedw Colliery, in 1884 for £130,000. Under the new owners, Cory Brothers of Cardiff, production of steam coal increased. When the collieries were later sold to the Powell Dyffryn Company, production increased by a phenomenal amount, producing 950 tons of best quality steam coal per day at its peak in 1928,with a workforce of 1,200 miners.

During the colliery's last full working year in 1962, under the management of the National Coal Board, over 95,000 tons of coal were produced, giving an average daily output of 260 tons. The depression in the coal industry forced the colliery to close.

In 1964 the site was cleared, and developed as an Industrial Estate for small and medium-sized factory units to meet the needs of local and incoming businesses. The estate is still in constant use despite the fluctuations of the economic climate.

Gelli Colliery in 1905. The Managers house is the square white building to the right of the chimney.

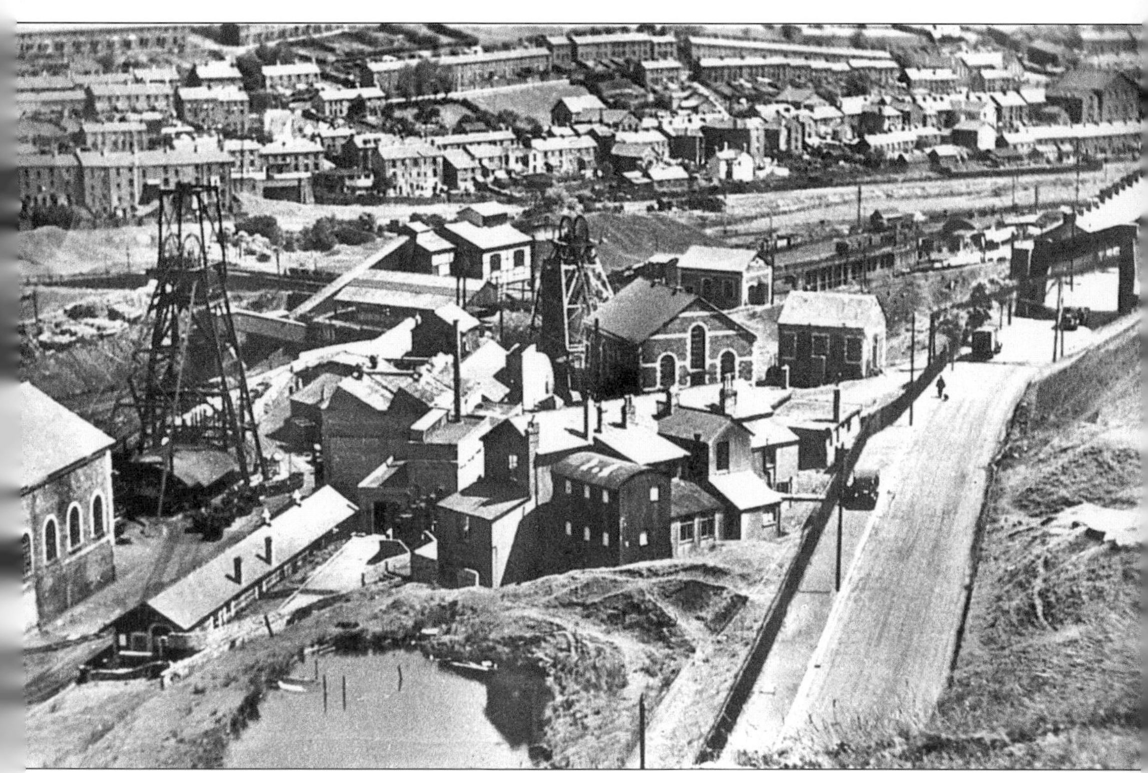

Gelli Colliery *c.*1935.

A recent view of the Gelli Colliery site, transformed with a variety of modern factories on the new estate.

Bodringallt Colliery, Ystrad

The colliery was sunk in 1864 by Warner Simpson and Company, produced more than 34,000 tons of coal, in its first year of operation.

In 1890, the colliery was purchased by David Davies and Sons, Ferndale, and subsequently became known as the Ferndale No.3 Colliery, later used by them as an air pit to improve ventilation at their other pits and the Nebo Level (closed in 1953) which was situated nearby. Towards the close of the nineteenth century the colliery was purchased by the Cory Brothers, Cardiff.

The fireclay found while working the Nos 2 and 3 Rhondda seams, and the great demand for bricks, led to a brickworks being set up at Bodringallt. This soon became the second largest producer of bricks in the Rhondda after Llwynypia. The bricks were used in the walling of shafts and in the construction of colliery buildings and houses. So great was demand that bricks were sent all over the South Wales coalfield.

The colliery was closed in 1959, at the same time as Nos1 and 5 pits at Ferndale. The site was cleared and then landscaped to blend in with the area.

A view of the Bodringallt Colliery, c.1920

The cleared and landscaped site of Bodringallt Colliery.

Pentre Colliery

In 1857 Edward Curteis of Llandaff, Cardiff leased the mineral rights of land at Pentre from Griffith Llewellyn of Baglan Port Talbot, and developed the Pentre and Church levels.

By 1864 shafts had been sunk to the deeper seams, and over 46,000 tons of coal were produced. In the same year Edward Curteis was joined by Messrs Greenhill and Ware and traded under the name of the Pentre Coal Company. The 4ft seam was struck at a depth of 198 yards, and a second pit was sunk.

In October 1869, the Pentre Colliery was purchased by Cory Brothers of Cardiff (ship-owners). They developed the colliery and between 1874 and 1884 almost trebled the annual output from 59,000 tons to 159,000 tons.

In 1871 there was a massive explosion in which a total of thirty-eight miners died. At the inquest, the verdict was that the explosion occurred because of a sudden discharge of gas, probably ignited at the furnace. No blame was attached to any of the officials connected with the colliery. The jury issued a strong recommendation that the suggestions of Mr Wales, the Government Inspector, should be carried out as soon as practicable, and that a night bankman should be constantly employed.

The colliery closed in 1929. By 1954 the area had been cleared of all colliery buildings and the site landscaped and prepared for playing fields and other recreational facilities.

Pentre Colliery, *c.*1906.

Another view of Pentre Colliery from the 1900s.

Pentre Colliery, Ystrad, 24 February 1871.

At 6.00 p.m. on Friday 24 February 1871, 150 workmen left the pit at the end of the day shift, and the night shift of thirty-six persons, both miners and hauliers, descended underground. At about 9.40 p.m. a violent shock was felt along the ground for more than a mile. This was immediately followed by a rush of ignited gas up the upcast shaft, destroying the mechanism required to raise and lower the trams. Extensive damage was done to the buildings adjoining the colliery, and some flying debris fell on houses several hundred yards away. Sulphuric fumes rising from the shaft prevented anyone approaching for some time.

A rescue party of six miners, including Mr William Rosser, the underground manager, descended into the pit by the downcast shaft. The explosion had resulted in the deaths of all thirty-six miners and also claimed the lives of two of the rescue team. Many of the bodies were never recovered, others were so badly burnt that identification was not possible.

At the time paperwork and records were sparsely kept, and after the explosion it was not possible to be completely sure who had entered the mine. Only twenty-four names were given out for those who died. In those days it seems the owners ran no disciplined checks, considering paperwork to be an unnecessary chore which did not create any profit!

Listed are the names of some of those who died:

Henry Backer, 36 years
Enoch Davies, 30 years
George Day, 36 years
John Griffiths, 22 years
Henry Haines, 17 years
John Hughes, 34 years
James Jones, 20 years
William Lewis, 38 years
John Michael, 28 years
Daniel Morgan, 24 years
William Rosser, 21 years (rescuer)
Joseph Thomas, 30 years (rescuer)
George Coburn, 32 years
Robert Davies, 23 years
Samuel Evans, 26 years
Thomas Griffiths, 48 years
William Howells, age unknown (rescuer)
George Jones, age unknown (rescuer)
Morgan Jones, 65 years
William Meredith, 17 years
John Mills, 36 years
David Morgan, 28 years
Joseph Sullivan, 28 years
Walter Williams, 35 years.

Maindy Colliery, Pentre

In 1864, David Davies of Llandinam, together with other partners, decided to exploit the mineral resources of the Rhondda Fawr and consequently leased mineral properties in Maindy and Cwmparc. Fifteen months later, having invested £38,000, there was no sign of any coal. He called a meeting of his workmen and informed them that as a result of his financial problems he could not go on any longer. His workmen, inspired with complete trust and loyalty, decided to continue working for one more week without payment. During that week, on Friday 9 March 1866, they struck the 4ft steam coal seam. It proved to be one of the finest quality seams in the coalfield.

The coal was shrewdly marketed under the title of Ocean Merthyr Coal. Few people had heard of the Rhondda Fawr, but Merthyr was known throughout the world!

By 1890, at the time of Davies death, 1,220 men were employed at the colliery, and more than 287,000 tons of coal had been raised. Production reached its peak in 1894, when 1,399 men were employed.

By 1931 only 215 men were employed, due to the economic depression, and when it was closed by the National Coal Board in April 1948, the colliery employed around 350 men. The coal tips were removed in the 1950s and the area cleared and landscaped. In the early 1990s a building programme for the construction of executive houses was implemented. Today it provides a first-class housing site which is the showpiece of the valley.

The colliery in 1910.

The colliery in 1922.

Monument erected on the site of the former colliery.

Tynybedw Colliery, Pentre

Sunk in 1876 by Edmund Thomas of Maindy Hall and George Griffiths of Pontypridd, on farmland leased from Crawshay Bailey of Nant-y-Glo, Tynybedw Colliery was popularly known as The Swamp.

It was sold to the Cory Brothers of Cardiff, together with the Gelli Colliery, in 1884, for the sum of £130,000. In 1913, it employed 660 men.

The colliery closed in 1933, when the Crawshay Bailey Estates would not grant more than a fourteen year extension to the lease. The site was cleared of all colliery buildings, the area flattened and some effort put in to blend it with the existing mountain slopes. It was then left so that in time the industrial scars would be covered by the growth of grass and ferns. A road was later put in by the Forestry Commission, who planted trees on the upper slopes of the mountain.

The colliery in 1905.

The colliery in 1913.

Cwmparc (Valley of Parkland)

Dare Colliery, Cwmparc.

Sunk in August 1866, by David Davies and Company, the Directors being David Davies and Thomas Webb, Llandinam; Morgan Joseph, Ystradfechan; John Osborne Riches, Aberdare; Abraham Howell, Welshpool; and Ezra Roberts, Tenby; the colliery was worked profitably by them until it was sold in 1877.

It was purchased by the Ocean Coal Company Ltd, which by 1890 employed 748 men and boys, producing more than 184,000 tons of coal per annum. However, although manpower was increased, geological problems dictated that higher output could not be consistently achieved.

In 1935, the Dare became part of the Powell Duffryn Associated Collieries Limited, which was one of the largest coal producing units in the United Kingdom. The entire enterprise was taken over by the National Coal Board in 1947.

Despite being absorbed into the neighbouring Park Colliery in 1955, to become one of the largest coal producing units in the South Wales coalfield, it was closed by the NCB in February 1966. The area was later cleared and landscaped to blend in with the contours of the mountains and the valley floor, giving it a natural and pleasant look. Since then, various buildings, such as a Community Hall and houses have been built on the site.

Dare Colliery in 1903.

The colliery in 1909.

The colliery site today. The chapel is visible in the background.

Park Colliery, Cwmparc

The colliery was sunk by David Davies and Company (see Dare Colliery) and later became part of the Ocean Coal Company Ltd. In 1929, the first pit head baths in the Rhondda was installed there, and officially opened by Lord Chelmsford, the Chairman of the Barry Dock and Railway Company. It was of great benefit to the miners, and became the forerunner for the other collieries in the valleys, slowly putting an end to the tub in front of the fire.

In 1935, Park Colliery became part of the group of the Powell Duffryn Associated Collieries Ltd. However, the takeover did not go smoothly. In October a massive protest in the form of an eight day stay down was held against the owners decision to employ members of the SWMIU, the Scab Union. Many of the miners who took part suffered permanent damage to their health.

The colliery was taken over by the National Coal Board in 1947. Output continued to rise after nationalisation, and in 1955, prior to its amalgamation with the Dare Colliery, 2,291 men and boys produced 360,000 tons of coal. Output continued to rise for a few more years:

1956: 2,192 men, output 414,000 tons.
1957: 2,136 men, output 423,000 tons.

Due to various production problems, output started to decline, resulting in a reduction of manpower at both collieries. The NCB closed the colliery (and the neighbouring Dare Colliery) in February 1966. The area was later cleared of all colliery equipment, and landscaped to blend in with the surrounding mountains.

It is interesting to note that, in 1903, the miners of both collieries, the Park and the Dare, decided to build the Park and Dare Workmans' Library and Institution, to provide recreational facilities for the miners and their families. The miners, through their Mineworkers Trade Association, pledged to pay one penny in every pound that they earned towards its construction and upkeep. Even today this magnificent building plays a very prominent part in the cultural life of the Rhondda.

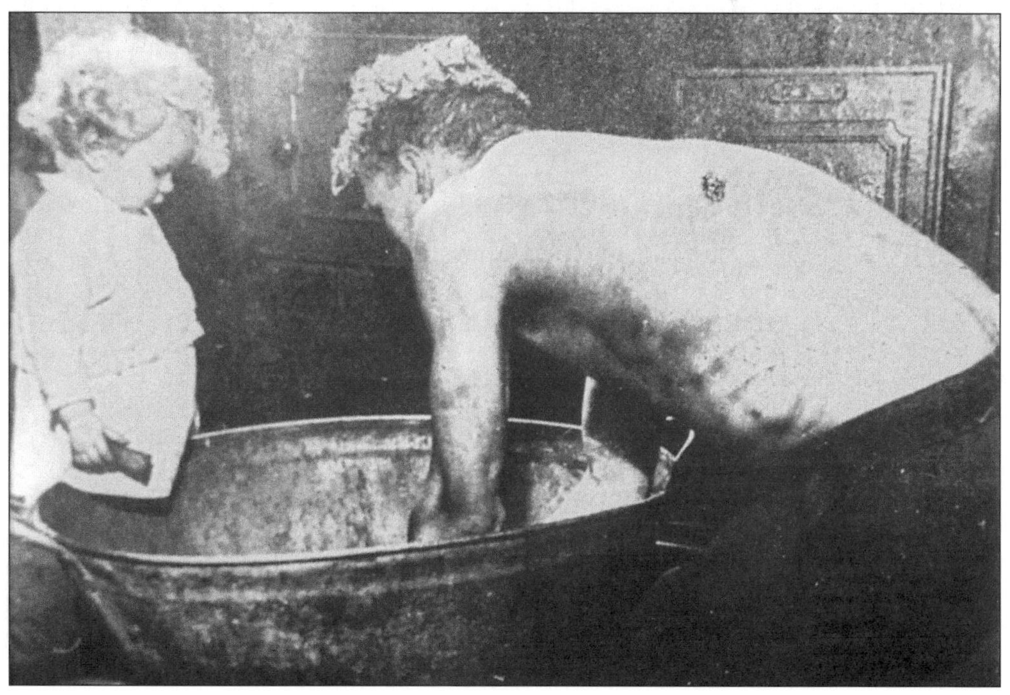

Park Pit, *c*.1920.

The landscaped area today. Note the colliery manager's house and the capped shaft.

Previous page: A typical scene in the early 1900s of bathing in front of the fire after finishing a shift in the local colliery. In a house with five or six working in the colliery, the father would bathe first, followed by his sons in descending age order. They would bathe their top half first, then their bottom half afterwards. With the water in such a state, would it be worth the youngest bathing at all?

Treorchy (Home of the Orchwy)

Abergorki Colliery, Treorchy

J.H. Insole of the Cymmer Colliery, Porth, purchased the Abergorki Level and plant in June 1862 for £7,000 from Houghty Huxham (former manager of the Bute Merthyr Colliery), and in 1863 he sank the deeper seams which commenced production in 1865.

The colliery was sold in 1874 to Burnyeat, Brown and Company from Liverpool and Whitehaven, who sank the shaft deeper into the steam coal seams and greatly increased output from 50,000 tons in 1866 to 220,000 tons in 1875. In 1913, 1,926 men were employed at the colliery, but by 1920 numbers had declined to 1,800 men.

In 1926 the colliery was purchased by the Ocean Coal Company, and by 1934 only 760 miners were employed. Coal at that time was being transported from the pit to the marshalling yards at Tylecoch for distribution. In order for the trams of coal to reach the yard they had to cross the main road, and in the interests of safety, gates were situated at the end of Maes y Ffrwd street, Treorchy, to control the traffic.

Due to geological problems, the colliery closed in April 1938, and after nationalisation the site was cleared of equipment and buildings, the shaft filled in and the site left to Mother Nature. Over a number of years the local authorities used the site for the disposal of refuse. The whole area was then covered in topsoil, and trees planted by the Forestry Commission, resulting in the pleasant surroundings of today.

The gates at Maes y Ffrwd, Treorchy, *c.*1900

The colliery site in the 1920s.

Tylecoch Colliery, Treorchy

In 1854, a Newcastle firm, Messrs John Carr, Morrison and Company, obtained the mineral concession, and a shaft was sunk to a depth of 154 yards, to the upper four foot seam. Many difficulties were encountered in working this seam, and in April 1858 the company was put up for sale, although it was not until December 1865 that it was purchased by Thomas Jones of Cardiff. Jones deepened the shaft to reach the lower four foot seam. During the sinking of the shaft, an explosion took place resulting in the deaths of Enoch Francis, Contractor, and a Mr. Bedlington, Mining Engineer.

In 1877 the colliery changed hands and was renamed the Tylecoch Coal Company. It was later resold to Thomas Jones of Cardiff, who worked it until 1883 when it was sold to two local colliery managers, Messrs Thomas and Evans, but after heavy financial losses it closed in 1885. In June of that year, the Phoenix Coal Company reopened the colliery, but after three years, the mortgagees took possession, and in 1891 the colliery was sold to the Marquis of Bute, and worked until its closure in 1895. In 1917 it was bought by the United National Company, who deepened the shaft with a view to becoming more productive.

In 1918 Tylecoch became part of the D.R. Llewellyn Combine, but little was gained by this, and the colliery remained closed. The colliery experienced production difficulties throughout its chequered life, and was the only colliery in the Rhondda which did not produce any financial profit for its successive owners, proof of their insatiable desire to exploit the natural resources of the area. The colliery site remained derelict for many years until the area was cleared and landscaped to construct a large secondary modern school on the site. Completed in 1965, when comprehensive education was introduced, the school did not become fully comprehensive until 1970. 1,875 pupils would attend the school, to be taught a range of subjects by an academic staff of over 100 men and women.

Tylecoch Colliery, c. 1920.

Treherbert (Named after the Herbert Family)

Ynysfaio Colliery, Ynyswen

The mineral property of a thousand acres had been purchased in 1854 by the Troederhiw Coal Company, and a shaft was sunk by James Thomas of Ynyshir and his partners, Mathew Cope of Cardiff and John Lewis of Aberdare. In 1859 the company sank the shafts deeper to mine the high quality steam coal seams, and a steady economic growth occurred. Subsequently Nos 1, 2 and 3 pits were deepened into the lower seams, resulting in a further increase in coal output. In 1874 output was approximately 36,000 tons, while in 1877 output had increased to approximately 102,000 tons. In 1888 the old Nos 1,2 and 3 pits employed 562 men, while in 1914 the new pits, Nos 3 and 4, employed 542 men.

In 1918 the colliery was still a private company, with chief patrons Mathew Cope and Sir W.J. Thomas Ll.D. They sold the property, comprising 1,000 acres, to Mr D.R. Llewellyn of Aberdare. The sale, taking effect from 1 January 1919, included the Tylecoch Colliery which had been closed in 1895.

Production from the colliery was approximately 300,000 tons per annum, which Llewellyn looked to increase when sinking Nos 3 and 4 shafts. However, either because of the economic depression, or because of geological problems, production declined rapidly and men were laid off. After nationalisation in 1947, only three men were employed at the colliery for pumping purposes.

The colliery site in 1909.

Ynysfaio Colliery in 1913.

Bute Collieries, Treherbert

In September 1855, Bute Merthyr Colliery, the first steam coal colliery, was sunk in the Rhondda on the Cwm-Saerbren farm by William Southern Clarke, Chief Mining Engineer to Lord Bute, for the Trustees of the Estate of Lord Bute.

On 21 December 1855, thirty-eight wagons of Rhondda steam coal was transported from Cwm-Saerbren by means of teams of horses as far as Gelligaled, and then loaded onto the first mineral train to Cardiff, where it was distributed as a Christmas gift to the poor.

In 1856, the Taff Vale Railway was extended to the Bute Merthyr Colliery, enabling coal to be loaded into wagons for transportation. The colliery continued to flourish as the railway yard was expanded. In around 1865 the new locomotive shed was built, incorporating a turn-table to position locomotives on any of the seven roads as required, for rapid transportation to Cardiff Docks. The building was commonly known as the Round Shed.

High demand for steam coal continued, and the Lady Margaret Colliery was sunk in 1877. The Blaenrhondda line was completed in 1894, and the Rhondda Tunnel opened in 1889, resulting in the building of a new yard and the existing station layout being rearranged. With the coupling of the Taff Vale Railway and the Swansea Bay Railway, coal could be quickly transported to the main docks of Cardiff and Swansea, and the site became the main terminus for coal transportation.

The Lady Margaret Colliery was closed in 1909, but retained for maintenance purposes, even after nationalisation in 1947. The Bute Merthyr Colliery, owned then by the United National Collieries, closed in 1926, with the loss of 722 jobs.

Treherbert Railway Station, *c*.1912. The locomotive turn-table can be seen to the right of the photograph.

Bute Merthyr Colliery, *c*.1910.

Lady Margaret Colliery, *c*.1906.

Tynewydd Colliery, Treherbert

Ebenezer Lewis, owner of the Bwllfa Dare Colliery, Aberdare, purchased his first lease of mineral property in the Rhondda Valley in 1865, at Tynewydd, opposite Cwm-Saerbren, on the east side of the River Rhondda.

An old level, which had previously worked the No.2 Rhondda seam before being abandoned, was immediately reopened, and a shaft sunk to the lower four and six foot seams. In 1866, more than 25,000 tons of good quality coal was raised.

In 1868, the colliery was purchased by the Rhondda Merthyr Colliery Co. for the sum of £50,000. Ebenezer Lewis was appointed Managing Director, and the site became widely known as the Rhondda Merthyr Colliery. It was profitable until 1879, when financial problems were caused by a serious geological fault, affecting production and involving the colliery in a legal dispute.

In 1887, the colliery was purchased by Messrs L. and H. Gueret, and once the initial geological problems were solved, efficient production was re-established. By 1910, it was producing over 100,000 tons of coal per annum, yet it closed in 1911.

The levelled and landscaped area became the site of the Treherbert Recreational Park in 1963, with a modern play area for children, and a large playing field used for many sporting events, including all the league matches of the Treherbert Rugby Football Club.

The park on the colliery site today, with the playing field a prominent feature.

Blaenwm (Top of the Valley)

Tydraw Colliery, Treherbert

The colliery was sunk by Thomas Joseph of Aberdare in 1865, on behalf of the Dunraven United Collieries Company Ltd, but the concern did not prosper, and went into liquidation in 1866.

In 1872 the colliery was sold for the sum of £155,000 to Edmund Hanney Watts, William Milburn, and Edward Stout and produced coal for many years before being sold to Cory Brothers of Cardiff. At the end of the First World War the colliery employed over 730 men.

Output diminished rapidly after nationalisation in 1947, due to geological and economic problems.

1954: 383 men, output 92,000 tons.
1956: 387 men, output 64,000 tons.
1958: 358 men, output 53,000 tons.

The colliery was closed by the National Coal Board in January 1959.

Many of the men who worked here and members of the local community referred to it as the Dunraven Colliery.

The site was later cleared and landscaped to blend in with the surrounding area. In 1996, members of the Rhondda Valley Railways' Society embarked on an ambitious project to build a £3.5 million theme park, including a Steam Railway Heritage Centre on the site of the old colliery.

Tydraw Colliery, *c*.1920.

Glenrhondda Colliery, Blaencwm

Popularly known in the Rhondda Valleys as the Hook and Eye, due to its reputation for improvisation, rather than replacement, Glenrhondda's operating costs were of paramount importance to the owners of this colliery.

The No.1 pit was opened in 1911 by the Glenavon Garw Colliery Company, and employed 230 men. The first coal to come out of the pit was taken from the No.1 Gorllwyn Seam. The No.2 pit was opened in 1921. Due to geological problems experienced in this area of the coalfield, output was not as high as in other collieries in the Rhondda. Although manpower was increased, output continued to fall. This can be seen from the statistics given below:

1954: 391 men, output 76,600 tons
1956: 394 men, output 110,000 tons
1958: 392 men, output 97,000 tons
1960: 403 men, output 82,700 tons
1961: 381 men, output 76,700 tons

The colliery was closed by the National Coal Board in September 1966. There were 321 men working then, and the majority were transferred to other pits in the locality. However, fifty-four men were retained at the colliery for demolition purposes. The site was dismantled and landscaped to blend in with the surrounding mountain area. It is now very popular for pony trekking. The only sign that there was ever a colliery in the area is the mound where the shaft was capped.

Glenrhondda Colliery in the 1960s.

Blaenrhondda (Source of the river Rhondda)

Fernhill Collieries

After the sinking of the Blaenrhondda Colliery in 1869, Ebenezer Lewis secured a mineral concession on land 200 yards from it, on behalf of the owners J. Marychurch and Partners. Nos.1 and 2 pits were sunk to the six foot seams in 1871, and sold later that year to three Yorkshire merchants, Messrs Crowley, John, and Oldroyd of Dewsbury. They sank Nos. 3, 4, and 5 pits in 1872, and in 1877 sold the whole concern to George Watkinson and Sons, owners of the Buckley Collieries, Flintshire. In 1893, George Watkinson and Sons bought the Blaenrhondda Colliery, which was then absorbed into the Fernhill combine, to form a company called Fernhill Collieries Limited.

After nationalization in 1947, manpower and output levels remained fairly consistent, with 1957 a boom year. However, geological problems eventually caused a decline, despite an increase in manpower.

1954: 934 men, output 230,000 tons
1957: 987 men, output 263,000 tons
1958: 987 men, output 217,000 tons
1961: 1,015 men, output 183,000 tons

In March 1966, the National Coal Board merged the colliery with Tower Colliery at Hirwaun, and diverted output there via underground workings. The Fernhill section of the mine was closed in July 1978. The NCB later demolished the site and cleared the old buildings before landscaping the area. It remained in this state until 1987, when permission was given by the then Rhondda Development Services Committee for development, by Western 7 Promotions of Birmingham, of a Western Holiday Theme Park. It would cost an estimated £1.0 million, and create more than fifty jobs. The first phase of the development would consist of a saloon with daily entertainment, a sheriff's office, a general store, livery stables, a gaol, a gun shop, a courthouse, a barber's shop, and a bank. Later there would be a Western Museum, an Indian village, a mansion house, and a reconstruction of the Alamo. In order that the development blend in with the environment the company promised to plant 1,000 trees while landscaping the area.

The grand opening day on 2 May 1987 proved a disaster, with many of the buildings promised in phase one either incomplete or unstarted. Machinery and building materials were scattered over the development, constituting a constant danger to visitors. It never recovered from this experience, and the company went into liquidation in June of that year.

In October 1987, a local business man took over the site, renamed it the Rainbow Valley Ranch, and refurbished the Golden Garter saloon at a cost of £65,000. Despite his attempts to make the area into a major development and tourism complex, the whole concept collapsed and the site became derelict. Fernhill Mining Ltd took over the site for coal reclamation and landscaping. In March 1998, the Wild West buildings were demolished.

Above and below: Fernhill Colliery, *c.* 1907.

Fernhill Colliery in the 1960s.

Blaenrhondda Collieries

The Blaenrhondda Colliery, (also known as the North Dunraven Pit), at the northern extremity of the Rhondda Fawr, was sunk in 1869 by George Locket and James Marychurch of Cardiff, and Herbert Kirkhouse and Rees Jones of Aberdare. The combine traded as The Cardiff and Merthyr Steam Coal Company.

In 1875, when the shaft was completed to the four foot seam, the colliery was sold to the London and South Wales Coal Company and worked as a going concern.

In 1893 the colliery was sold to George Wilkinson and Sons, who were the owners of the Fernhill Collieries. Blaenrhondda was then amalgamated into the Fernhill concern. Output from this and the other collieries was high and in some instances reached as much as 600 tons per day. Being part of the same complex as Fernhill, it closed at the same time in March 1996.

Blaenrhondda Colliery, 1911.

Six

Rhondda Fach

Striking miners marching to Pontypridd Court in 1910.

Ynyshir (Long Island or Meadow)

Lady Lewis Colliery, Ynyshir

This colliery opened in 1904, owned by Lewis Merthyr Consolidated Collieries Limited. It was worked in conjunction with the Lewis Merthyr Colliery at Trehafod, and owned by the same company. Its development was rapid, and by 1913 employed 1,228 men.

Coal production continued for many years. Lady Lewis Colliery ceased production in 1926, and was then used for ventilation and pumping purposes for the Trehafod pits. After nationalisation in 1947 it continued to be used for ventilation and pumping by the National Coal Board, while the site itself was used for the stockpiling of surplus coal.

In 1967, the local council was successful in its negotiations with the NCB, obtaining the eight and a half acre site for the erection of factory units, of various sizes, to bring desperately needed employment to the area. The site was cleared and advanced factories were built by the Welsh Industrial Estates Corporation. Today they provide employment to many local people.

Lady Lewis Collieries, *c.*1910.

Another view of the Lady Lewis Colliery from the 1920s.

Ynyshir Colliery

Ynyshir was sunk in 1845 by Messrs Shepard and Evans, trading as the Ynyshir Coal Company. In 1856 Francis Crawshay, who had become the sole owner of the Treforest Tin Works, purchased the colliery, which he extended in order to supply coal to his Tin Works. At that time approximately 100 men and boys were employed at Ynyshir Colliery. In 1873 it was purchased by Thomas Jones of Maendy House Ynyshir, who changed its name to Jones Navigational Colliery. He sank the shaft deeper in order to work the No.2 Rhondda seam, which he struck at a depth of 200ft.

The colliery closed in June 1909 because it was not profitable to work. The site was cleared and today is the local football field.

A view of 'Jones' Pit', Ynyshir from 1904.

Standard Collieries, Ynyshir

In 1874 William James Thomas, with Daniel Thomas of Dinas, began mining steam coal in the Rhondda Fach, leasing a considerable tract of land. He began by sinking two shafts, which became known as the Standard Collieries. In 1877, the four foot seam was struck in the No.1 pit at a depth of 377 yards; and in 1878, the same four foot seam was struck in the No.2 Pit, at a depth of 525 yards. Combined production reached over 188,000 tons that year.

Coal from this colliery was highly esteemed by the British Navy, and was in great demand in many foreign countries, including France. W.J. Thomas built a large house, in its own grounds, opposite the colliery, called Brynawel. From here he conducted all the colliery affairs, even working out the wages of the miners.

In 1914, the Standard Collieries were amalgamated with the United National Collieries Company Limited. They employed over 1,300 men, with coal production reaching over 360,000 tons. The combine later became part of the Ocean Coal Company. The colliery was never worked by the NCB after nationalisation in 1947, but remained open for ventilation and pumping purposes for the neighbouring collieries.

In later years, the colliery was dismantled, the shafts capped, and the land cleared and landscaped in preparation for small factory units. Today, while there are some small factory units in use, the site is mainly dominated by the Wilpak factory, which was built in 1966. In April 1985, during maintenance work at the factory, it was found that the capped shaft located inside the factory had collapsed as filling material washed away over the years. The 800ft shaft was filled in with 5,000 tons of hardcore and the final 100ft capped with concrete. The work cost around £60,000, paid for by Rhondda Borough Council, who purchased the land in 1965 for the sum of £1,750. In 1998, Wilpak Ltd became part of a multi-international group of companies, and changed their name to Danisco Pack Rhondda.

YNISHIR

STEAM COAL CO.

W. JAMES THOMAS, Proprietor.

OWNERS AND SHIPPERS OF

STANDARD MERTHYR

STEAM COAL.

ON THE BRITISH ADMIRALTY LIST.

* * *

Used by the Navies of Foreign Governments, and by the chief Steamship Lines of the World.

USED BY THE CHIEF

ENGLISH AND CONTINENTAL RAILWAYS.

HEAD OFFICE:
EXCHANGE, CARDIFF.

Telegraphic Address:
"STANDARD, CARDIFF."

A typical advertisement issued from a local colliery.

Above and below: Two views of the Standard Collieries, *c*.1904.

Wattstown (Named after James Watt)
National Colliery, Wattstown

The mineral rights were obtained for Pont Rhyd-y-cwch, later to be known as Wattstown, by Crawshay and William Partridge Bailey in 1881. The same year, three Rhondda coal owners, Ebenezer Lewis, Henry Lewis, and Mathew Cope formed the United National Coal Company, and sank the Nos1 and 2 shafts to a depth of 450 yards, in order to work the six foot seam. The two sites were known locally as Pwll Mawr, and Pwll Bach. These were later worked by other companies, such as Messrs Griffiths and Company; and Messrs Watts and Company.

A terrible explosion occurred on 18 February 1887, in which thirty-nine men died, and at the inquest which followed, the jury found that Griffith Griffiths died after an explosion of gas at the National Colliery, in the parish of Llanwonno, at 6.40 p.m. on 18 February 1887, and that they could not, in consequence of the lack of evidence, find out the cause of the explosion nor where it started. It was the unanimous opinion of the jury that the management of the said colliery was conducted in a loose manner, and not so carefully as the case required. The same verdict applied to thirty-seven more of those killed. With regard to the thirty-ninth, the following verdict was recorded: that Albert Shewry met his death from accidentally falling down the upcast of the of the National Colliery on the 18th. February 1887, and was found dead in the sump on the 23 February 1887.

The colliery was later sold to the National Steam Coal Company Limited, under the management of Mr. Henry Lewis. Although its name was later changed to the National Collieries Company Limited, it was still known locally as Cwtch Colliery.

In 1894 the shafts were deepened to work the richer lower five foot steam coal seams. On the 11 July 1905, a massive explosion occurred which claimed the lives of 119 men and boys. The mining communities of Pontygwaith, Wattstown, and Ynyshir were devastated. Such was the scale of the disaster, that on the day of the mass funeral the corteges were so long that when the last hearse left Wattstown, the lead hearse was entering the gates of Llethr-Ddu Cemetery, Trealaw, nearly three miles away. This was the worst mining disaster in the valleys since Ferndale in November 1867. The official cause of the tragedy was shot firing igniting the newly released gas in the coal barrier, at the cross heading between the sinking pit and the upcast shaft. The official recommendation was that shot firing should be absolutely prohibited except between shifts and only shotmen should be in the pit at the time. In addition a thorough system of watering the roadways, sides and roofs should be compulsory, and the authorities should see that the existing regulations for issue and use of explosives be stringently observed.

In 1914, the colliery was amalgamated with Standard Collieries. The combine later became part of the Ocean Coal Company. Output continued to rise, and in 1947, after nationalisation, output for the year was 161,000 tons, with 748 men and boys being employed at the colliery. Although there was an increase in manpower in later years, geological problems caused output to fall, and the colliery was closed in November 1968. Shown below are some statistics of manpower and output in the latter years of the colliery:

1958: 720 men, output 212,000 tons
1959: 633 men, output 197,000 tons
1960: 585 men, output 203,000 tons
1963: 540 men, output 128,000 tons
1966: 605 men, output 123,000 tons.

National Colliery, Wattstown, 18 February 1887

At about 6.30 p.m. on Friday 18 February 1887 a terrific explosion occurred at the colliery, resulting in the deaths of thirty-nine miners. The majority of the day shift workers had been brought out of the pit, but due to a delay at the surface, caused through problems on the winding gear, the night shift had not been able to go down the pit, otherwise the death toll would probably have been greater. Thomas Griffiths and William Read survived the explosion but died later due to their injuries.

The names of those who died are:

Richard Arthur, 60 years
Frank Belbin, 26 years
Walter Bevan, 30 years
Richard Bryant, 24 years
John Charles, 22 years
Oliver Clements, 23 years
John Curley, 36 years
Daniel Davies, 22 years
Henry Davies, 32 years
Morgan Davies, 40 years
Morris Davies, 28 years
David Edwards, 24 years
John Evans, 34 years
Morgan Gibbon, 26 years
Edward Goodwin, 25 years
Griffith Griffiths, 51 years
Thomas Griffiths, 25 years
William Guy, 18 years
Philip Jeffries, 28 years
Daniel Jones, 52 years
John Jones, 36 years
Thomas Jones, 23 years
Watkin Jones, 23 years
John Lewis, 32 years
William Llewellyn, 31 years
Thomas Morgan, 38 years
David Owen, 33 years
Richard Powell, 37 years
John Price, 29 years
Richard Pritchard, 42 years
William Reed, 22 years
Samuel Richards, 27 years
William Roberts, 22 years
Albert Shewry, 24 years
Thomas Tovey, 28 years
Thomas Tudor, 38 years
Charles Williams, 25 years
Henry Willford, 32 years
William D. Williams, 14 years.

National Colliery, Wattstown, 11 July 1905

On Tuesday 11 July 1905, at about 11.45 a.m., an explosion occurred which resulted in an appalling loss of life. Newspapers described the explosion as a lurid flash that leapt out of the main shaft, and a thunderous roar that shook the whole of the surrounding area. The colliery was covered in smoke, and a sulphurous stench enveloped the area. Fragments of debris which had shot out of the shaft after the explosion, fell on the buildings which surrounded the colliery.

At the time of the explosion 930 men were employed at the No.1 pit, and 120 men were employed at the No.2 pit. Of these 117 were killed immediately, either through the force of the explosion or the effects of the afterdamp which occurred. Two survivors were found alive, John Reeves who died six hours later, and John Dando who died fifteen hours later. This brought the total death toll to 119. The explosion was caused by shot firing in the coal barrier at the cross heading between the sinking pit and the upcast shaft. These are the names of those who died:

James Baines, 55 years
Emmanual Beard, 14 years
Robert Billett, 34 years
George Chidgey, 18 years
John Clancey, 38 years
John Dando, 27 years
Charles Davies, 18 years
David Davies, 37 years
David G. Davies, 23 years
David Davies, 47 years
Isaac Davies, 29 years
Joseph E. Davies, 18 years
Thomas Davies, 38 years
William Eastment, 45 years
Thomas Edwards, 23 years
J.Evans, 29 years

W.H. Evans, 17 years
Fred Fletcher, 25 years
James Gibbon, 48 years
William Basset, 14 years
Ammon Billett, 22 years
Samuel Bird, 16 years
Charles Clancey, 17 years
Robert Cross, 38 years
William Daniel, 40 years
Charles H. Davies, 15 years
David G. Davies, 26 years
David Davies, 32 years
Enoch Davies, 21 years
John Davies, 26 years
J.H. Davies, 14 years
Thomas Oavies, 40 years
Gwilym Edmunds, 19 years
George Evans, 26 years
Morgan R.Evans, 25 years
William J.Evans, 18 years
Thomas Flower, 41 years
John Gibbon, 25 years
Thomas Gibbon, 15 years
Robert Hallett, 38 years
James Healing, 14 years
Thomas Howells, 15 years
William E. Hudd, 16 years
William T. John, 23 years
David Johnson, 16 years
John Jones, 36 years
Thomas Jones, 15 years
Arthur Kemp, 43 years
Thomas King, 47 years
Ben Lewis, 16 years
William H. Lloyd, 16 years
Albert Marshall, 33 years
William Meredith, 55 years
John Morgan, 59 years
Richard Morgan, 18 years
William Morgan, 17 years
David T. Morris, 15 years
Charles E. Perry, 19 years
George Perryman, 38 years
William Perryman, 14 years
David Powell, 32 years
Oliver Pritchard, 38 years
Thomas J. Prosser, 16 years
David J. Rees, 15 years
John Rees, 17 years
John Rees, 37 years
Alfred Richards, 46 years
Elias Roberts, 25 years
Samuel Smith, 14 years
Morgan J. Stuart, 15 years
John Tingle, 14 years
Alfred Uzzel, 40 years
Benjamin Walters, 37 years
David Williams, 26 years
Idris Williams, 21 years
John M. Williams, 14 years
Thomas Williams, 15 years
Frank Wiltshire, 14 years
R. Yell, 20 years
W. H. Goldsworthy, 20 years
William Hallett, 17 years
John Howells, 42 years
Samuel Hughes, 21 years
Evan John, 18 years
David Johnson, 64 years
Isaac Jones, 17 years
Thomas Jones, 51 years
William Jones, 60 years
Alfred King, 22 years
Benjamin Lewis, 53 years
Thomas Lillicrop, 25 years
Flurence Mahoney, 17 years
Samuel Mason, 34 years
Edward Morgan, 16 years
John Morgan, 25 years
William Morgan, 15 years
William J. Morley, 27 years
Thomas Owen, 33 years
C.E. Perry, 39 years
Thomas Perryman, 19 years
David Phillips, 26 years
D.S. Powel1, 15 years
John H. Probert, 20 years
David Rees, 38 years
David J. Rees, 15 years
John Rees, 50 years
John Reeves, 19 years
William Richards, 14 years
Edward J. Sampson, 17 years
Thomas H. Smith, 17 years
William M. Thomas, 17 years
John Turberville, 14 years
John Uzzel, 15 years
John R. Walters, 15 years
Edwin C. Williams, 16 years
John Williams, 44 years
Richard Williams, 43 years
W. Albert Williams, 13 years
F. Woodham, 33 years

The funeral procession of many of the victims of the Wattstown disaster. Such was the length of the procession that as the lead hearse was entering the gates of Llethr-Ddu Cemetery at Trealaw, the last was just leaving Wattstown.

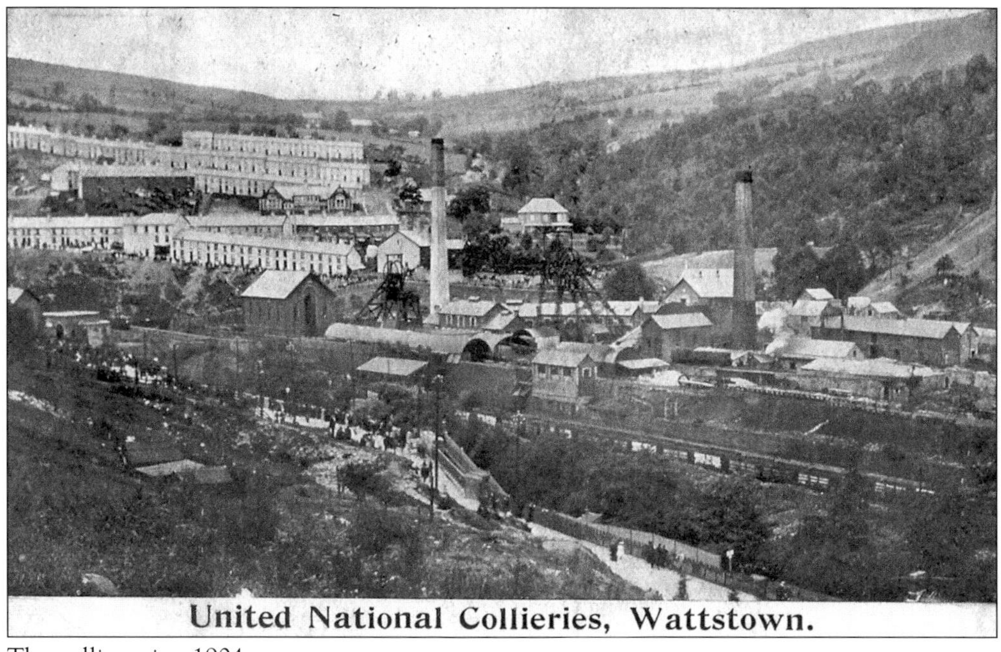

United National Collieries, Wattstown.

The colliery site, 1904.

SCENE OF THE WATTSTOWN DISASTER—NATIONAL COLLIERY.

Scene of the disaster, 1905.

Tylorstown (Named after Alfred Tylor)

The Tylorstown Colliery, Tylorstown

In 1858, Thomas Wayne sank a shaft in the area known today as Tylorstown. The site, known locally as Pwll Waynes (Waynes Pit), was officially called Cynllwyn-du, but was later renamed the Pontygwaith.

In 1872, Alfred Tyler, after whom Tylorstown was named, bought the mineral rights of Penrhys, and sank two shafts, known as the Penrhys Colliery. This was purchased in 1894 by David Davies and Company, who in 1896 also purchased the Pontygwaith Colliery. David Davies subsequently renamed them the Nos 6 and 7 (Penrhys), and No.8 (Pontygwaith) Collieries.

On 27 January 1896 a terrible explosion took place at the collieries, and fifty-seven miners lost their lives.

At the inquest the jury were of the opinion that the cause of the explosion was the firing of a shot in gas in the Daniel Williams headings of No.8 pit. The air passing through the faces was charged with gas, and the shot came in contact with the film of gas in the roadway. Coal dust accelerated the explosion. The jury believed that no one living was responsible for the explosion.

Under Davies's ownership the three collieries were connected together to improve ventilation. At that time upwards of 1,700 men and boys were employed, producing a daily output of 1,800 tons of high grade steam coal. In 1901, David Davies and Company sank the shaft for what was to become the No.9 Colliery. In 1908, David Davies undertook the electrification of the collieries, together with the Ferndale Collieries. The electricity was supplied from a new Power Station, which was built opposite the No.8 and 9 pits at Tylorstown, and at the end of the year, all haulage, ventilation, pumping, and screening plants were electrically driven. Unfortunately the Power Station was gutted by fire on the 24 May 1924.

In their heyday the collieries gave employment to around 3,000 men and boys. Yet production and geological problems arose, and the No.6 and 7 pits closed in 1936, retained only for pumping and ventilation purposes. The No.8 pit also closed in 1936, while the No.9 continued production. It was taken over in by the National Coal Board in 1947 and, despite a brief surge in production, finally closed in October 1960.

The whole site remained derelict for a number of years, until in 1980 a £1.2m land reclamation scheme was prepared. Implemented in 1981, the scheme diverted the river and removed over half a million tons of waste spoil to create fourteen acres of land for future development. The site of the Nos 8 and 9 collieries was laid out and landscaped to accommodate a football pitch, play area and health centre for the benefit of the local community. However, it was not until 1993 or 1994 that the Rhondda Fach Sports Centre was built on the site of the Nos 6 and 7 collieries. The remainder of the site was flattened and landscaped.

Nos 8 and 9 collieries in 1916.

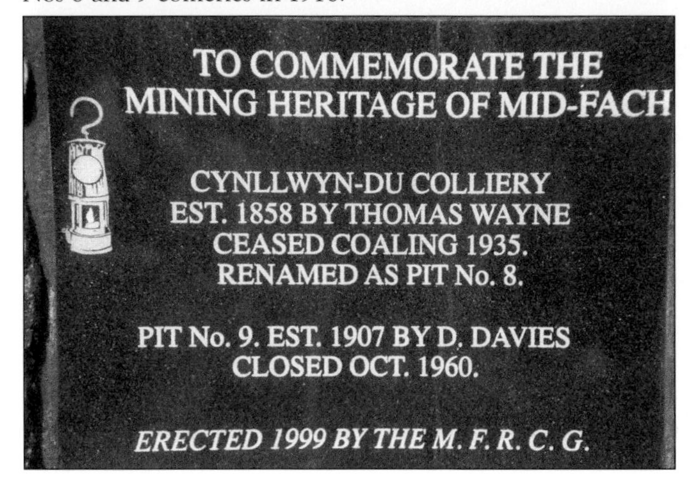

Memorial to the memory of the miners who worked and died at the now demolished Tylerstown Nos 8 and 9 collieries. This memorial was erected by the Mid Fach River Care Group.

View of Tylerstown Colliery from Stanleytown, *c.*1900

A general view of Tylorstown.

Tylorstown Colliery, Tylorstown, 27 January 1896

At about 5.30a.m. on 27 January 1896 an explosion ripped through the three shafts of the Nos 6, 7 and 8 pits, causing the deaths of fifty-seven miners. Such was the ferocity of the explosion that it not only caused numerous roof falls in all districts, but put the ventilation system of the colliery out of action, resulting in the build up of sulphurous and other noxious fumes, which caused the death of many of the victims through suffocation.

The names of those who died are:

Walter Barrett, 17 years
John Bowen, 34 years
Abraham Charles, 29 years
John Collins, 27 years
Henry Davies, 25 years
Sidney Davies, 37 years
Richard Davies, 34 years
Solomon Davies, 25 years
David Davies, 52 years
Thomas Davies, 18 years
James Edwards, 25 years
James Evans, 38 years
Jesse Evans, 20 years
Richard D. Evans, 28 years
Richard Evans, 48 years
George Gardiner, 26 years
George Groves, 19 years
Thomas Hall, 17 years
David Harris, 49 years
Henry Harris, 18 years
Alfred Jackson, 15 years
James Jackson, 25 years
Griffith Jenkins, 35 years
William R. Jenkins, 36 years
David Jones, 27 years
George Jones, 19 years
Gwilym Jones, 29 years
Jacob E. Jones, 18 years
David Jones, 22 years
David Lewis, 22 years
George Lewis, 50 years
John Lewis, 40 years
Evan Morgan, 34 years
Samuel Morgan, 21 years
Charles Norman, 22 years
Albert Olding, 20 years
John Pearce, 64 years
Benjamin Phillips, 52 years
Isaac Pride, 19 years
Amos Pritchard, 46 years
Daniel Reardon, 24 years
David Rosser, 41 years
John Rowlands, 21 years
Robert Saunders, 30 years
Thomas Scourfield, 32 years
George Stapleton, 54 years
James Sutton, 38 years
John Thomas, 37 years
Richard Thomas, 35 years
John Watkins, 33 years
David Williams, 22 years
Lewis Williams, 19 years
Joseph Williams, 20 years
Samuel Williams, 32 years
Edwin Williams, 20 years
Henry Wiltshire, 17 years.

Memorial Stone for the demolished Tylorstown Nos 6 and 7 collieries. Situated on the riverbank, this stone was unveiled on Friday 1 May 1998, in the presence of many local dignitaries. It was erected by the Mid Fach River Care Group.

Ferndale (Glynrhedynog)

Ferndale Collieries, Ferndale

The No.1 shaft was sunk by David Davies and Son in 1862, to the four foot seam at a depth of 280 yards, later deepened to 400 yards.

On 8 November 1867, a terrible explosion occurred at the colliery, with the loss of 178 lives. Nearly every household within the vicinity of the colliery was affected. The official verdict was accidental death, with the management censured for not having installed an adequate ventilation system. Sadly, just as people were beginning to come to terms with the tragedy, disaster struck again.

Two years later on 10 June 1869, another explosion took place in which fifty-three men and boys were killed. The official verdict found that the deaths resulted from an explosion of firedamp at the colliery, but there was not sufficient evidence to locate where it happened. In the final summing up the jury regretted that the suggestions made by the jury after the 1867 disaster had not been fully carried out by the Manager and other officials, who put profit before safety.

In 1870, after the two explosions, Lewis Davies, son of David Davies, sank the No.2 shaft to improve ventilation at the No.1 pit. The No.4 shaft was sunk by David Davies and Son in 1876, three-quarters of a mile further north than No.2. In 1889 the No.5 shaft, known locally as the Blaenllechau, was sunk half a mile from Blaenllechau village. In latter years it was worked in conjunction with the No. 1 pit. The Ferndale No.3 was sunk in Rhondda Fawr in 1864. Known locally as the Boedringallt, it was purchased by David Davies in 1890.

The Ferndale Collieries were taken over by the Welsh Associated Collieries in 1927. The Nos 2 and 4 pits were closed in the early 1930s, and the colliery was taken over by the Powell Duffryn Associated Collieries in 1935, before passing to the National Coal Board in 1947. From small beginnings in 1862, the colliery had grown to become one of the largest suppliers of good quality steam coal in the South Wales coalfield.

1954: 585 men, output 146,000 tons
1955: 583 men, output 150,000 tons
1956: 568 men, output 126,000 tons
1958: 500 men, output 124,000 tons

The colliery (Nos 1 and 5) was closed by the National Coal Board in 1959. In the 1980s the colliery site and the well known Banana Tip was removed and the area prepared for a Leisure and Business Park, costing £2 million.

In 1986, on the site of the No.1 pit, construction work began on the building of a Memorial Park, including a coal tram filled with coal and mounted on a stone plinth. Buried in this plinth was a steel capsule containing the names of all the miners who had died in the terrible disasters of 1869 and 1876. The official unveiling of the monument and the opening of the Memorial Park took place in 1988.

In October 1990, the Oaklands Business Park was officially opened by the thenSecretary of State for Wales, the Right Honourable David Hunt MP. He stated that he hoped this would provide much needed employment for the area. Plans for the Leisure Park, however, provoked controversy. An outside consortium wanted to build an up-to-date sports stadium with modern facilities and a concert hall and restaurant/bar complex to draw in people from the valleys. The local community favoured a simpler layout to cater for their needs. After a number of years common sense prevailed and the local community had a park built to suit the young people of the area.

In 1991, the Greenwood Park was officially opened by Cllr Margaret Chard, Mayor of the Rhondda, accompanied by Cllr Kate Rees, Chair of the Recreation and Amenities Committee.

Ferndale Collieries, 1911.

The caption of the postcard should read : Nos2 and 4 pits, Ferndale. No.3 is in fact situated in the Rhondda Fawr.

The memorial in Greenwood Park. In the 1980s the reclamation of the colliery site began. In 1986 a memorial garden containing a tram of coal was built on the site of the No.1 colliery. This was in memory of the miners who died whilst working at this colliery, including those who perished in the disasters of 1869 and 1876.

Ferndale Colliery, Ferndale, 8 November 1867

On Friday 8 November 1867 at about 1.30 p.m., two blasts in rapid succession gave the first indication that a massive explosion had taken place. Although there was not much damage on the surface, below ground was devastated. The explosion shattered roadways, filling them with debris that caused choke damp by preventing ventilation. Many died of suffocation. A staggering 178 miners lost their lives, which at the time was the largest number ever to be killed in one incident in the district. At the inquest it was stated that, of the total 178 miners killed, 145 were burnt, five died from the violence of the blast, and twenty-eight suffocated. Due to a lack of communication in those days, the names of many of the victims were not published.

The names of those who died are:

John Atkins, 27 years
Joseph Brickley, 21 years
Daniel Burke, 22 years
Richard Burke, 24 years
George Cooper, 27 years
David Davies, 25 years
David Davies, 15 years
John Davies, 35 years
John Davies, 21 years
John Davies, 25 years
Joshua Davies, 22 years
Thomas Davies, 26 years
William Davies, 19 years
William Davies, 28 years
John Davis, 32 years
Richard Davis, 18 years
James Driver, 13 years
John Driver, 37 years
George Edwards, 15 years
John Edwards, 59 years
John Edwards, 32 years
Thomas Edwards, 38 years
David Evans, 17 years
David Evans, 23 years
Henry Evans, 19 years
Isaac Evans, 22 years
Joseph Evans, 35 years
William Evans, 47 years
Edward Griffiths, 19 years
John Griffiths, 17 years
Morgan Griffiths, 41 years
Morgan Griffiths, 16 years
Thomas Griffiths, 19 years
William Griffiths, 23 years
William Griffiths, 19 years
William Hammond, 25 years
John Hancock, 25 years
John Harris, 26 years
Horatio Hawkins, 19 years
Joseph Howlett, 33 years
Miles Hughes, 13 years
Robert Hughes, 13 years
Daniel Humphreys, 27 years
Evan James , 24 years
John James, 19 years
John James, 25 years
Jenkin Jenkins, 40 years
John Jenkins, 48 years
Charles Jones, 22 years
David Jones, 14 years

David Jones, 36 years
David Jones, 33 years
Evan Jones, 14 years
Evan Jones, 33 years
John Jones, 27 years
Morgan Jones, 36 years
Robert Lapthorne, 32 years
Benjamin Lewis, 20 years
David Lewis, 33 years
Evan Lewis, 21 years
Henry Lewis, 28 years
John Lewis, 28 years
John Lewis, 32 years
John Lewis, 30 years
Lewis Lewis, 38 years
Thomas Lewis, 19 years
William J. Lewis, 48 years
Llewellyn Llewellyn, 28 years
Meredith Llewellyn, 20 years
Edwin Lloyd, 30 years
John Lukey, 25 years
Benjamin May, 19 years
Evan Meredith, 29 years
William Miles, 21 years
Thomas Miller, 25 years
John Morgan, 22 years
John Morgan, 22 years
Peter Morgan, 47 years
Roger Morgan, 16 years
Thomas Morgan, 35 years
Ben Morris, 50 years
Benjamin Morris, 29 years
Caleb Morris, 41 years
Daniel Morris, 16 years
David Morris, 18 years
Ebenezer Morris, 49 years
Henry Morris, 27 years
John Morris, 21 years
John Morris, 15 years
Edward Moseley, 26 years
John Neath, 19 years
David Nicholas, 35 years
Thomas Nicholas, 18 years
William Nicholas, 15 years
Charles Owen, 47 years
John Owen, 25 years
John S. Owen, 23 years
Owen Owen, 17 years
Ben T. Parfitt, 16 years
Thomas Parfitt, 33 years
William Parker, 28 years
John Pascoe, 20 years
John Powell, 44 years
Thomas Powell, 13 years
Thomas Powell, 14 years
James Prosser, 19 years
John Prosser, 35 years
Benjamin Rees, 22 years
Henry Rees, 30 years
William Rees, 42 years
John Richards, 14 years
Thomas A. Richards, 22 years
Nathaniel Roach, 35 years
Evan Roberts, 25 years
Robert W. Roberts, 12 years
James Roblin, 51 years
Daniel Rowe, 15 years
Evan Samuel, 21 years
Benjamin Saunders, 20 years
David Stephens, 36 years
John Swanscott, 57 years
David Thomas, 22 years
David Thomas, 22 years
Isaac Thomas, 33 years
John Thomas, 16 years
Lewis Thomas, 13 years
Thomas Thomas, 48 years
Thomas Thomas, 39 years
Thomas Thomas, 25 years
William Thomas, 31 years
Charles Truscott, 19 years
Thomas Vaughan, 23 years
John Walters, 28 years
William Walters, 14 years
William Watkins, 22 years
William Wells, 13 years
David Williams, age unknown
David Williams, 28 years
Edward Williams, 15 years
Edward Williams, 19 years
Henry Williams, 19 years
Henry Williams, 28 years
Howell Williams, 17 years
Jenkin Williams, 19 years
John Williams, 50 years
John Williams, 25 years
John Williams, 23 years
Thomas Williams, 35 years
Thomas Williams, 54 years
William Williams, 17 years
William Williams, 29 years
William Williams, 30 years
William Williams, 31 years.

Ferndale Colliery, Ferndale, 10 June 1869

At 6.30a.m. on 10 June 1869, about 600 miners descended into the pit. At around 7.00a.m. an explosion occurred, sending earth tremors through the colliery, with dense clouds of coloured smoke and shooting flames from the shaft. The buildings were showered with light debris. A total of fifty-three men and boys died. Their names are:

George Allen, 33 years
George Chamberlain, 29 years
Daniel Davies, age unknown
Samuel Davies, 28 years
John Davies, 21 years
Edmund Davies, 23 years
Abel Dore, 24 years
John Edwards, 32 years
Thomas Elliot, 40 years
Henry Excell, age unknown
Joseph Excell, age unknown
James Griffiths, age unknown
William Hughes, age unknown
Richard Jones, age unknown
Morgan Jones, age unknown
Thomas Jones, 24 years
James Jones, 50 years
John Jones, 21 years
David Jones, 24 years
Morgan Jones, 13 years
Morgan Jones, age unknown
Thomas Jones, 28 years
Thomas Jones, 14 years
William Jones, 30 years
David Jones, 24 years
William Llewellyn, 17 years
William Mathews, age unknown
John Mathews, 28 years
Abraham Mathews, 14 years
Jacob Mathews, 48 years
John Morgan, 22 years
James Morgan, 34 years
James Morris, age unknown
Richard Morris, 23 years
James Nash, 32 years
Evan Phillips, 43 years
Thomas Price, age unknown
John Price, 50 years
Evan Richards, 48 years
Windsor Roderick, 23 years
David Samuel, 59 years
Samuel Straide, age unknown
Charles Taylor, 43 years
George B. Thomas, 31 years
Thomas Thomas, 34 years
Thomas Thomas, 27 years
Thomas Thomas, 22 years
John Thomas, 23 years
Edward Waddleton, 30 years
Bendigo Weeks, 24 years
James Wilkins, 13 years
Evan Williams, 25 years
David Williams, 21 years.

The site of the No.5 pit (Blaenllechau) today.

Maerdy (House of Steward)

Mardy Colliery, Maerdy

The Mardy Collieries, at the far end of the Rhondda Fach, were sited about 1,000ft above sea level. In 1873 Mordicai Jones, a native of Brecon (and Chairman of the Brecon Gas Works), bought the mineral rights of the Maerdy estate. In partnership with another Brecon man, Wheatly Cobb, the sinking of the Mardy Nos 1 and 2 pits commenced in 1875, reaching the first seam, the Abergorki, in December 1876. In 1878, the Nos 1 and 2 pits were deepened to reach the steam coal, in the two foot nine, the four foot, and the six foot seams. In 1879, the lease of the colliery was sold to the Lockett-Merthyr Company, under whose management the enterprise prospered. However, a high price was paid for such success, as on 23 December 1885 a terrific explosion occurred underground, resulting in the death of eighty-one miners. The official verdict concluded that an explosion of gas occurred in the Rhondda (West) district of the colliery. Where the gas ignited is unknown, but the jury were satisfied that it was not the result of shot firing in the hard heading.

The underground workings of the Nos1 and 2 pits were extended, and it was decided to sink another shaft to be known as the Mardy No.3 in 1893, about a mile further up the valley from the Nos1 and 2. In 1914 the Mardy No.4 was sunk in close proximity to the No.3.

The Nos1 and 2 pits closed in 1932, after being taken over by the Bwllfa and Cwmaman Collieries Limited. In 1935 the colliery was taken over by the Powell Duffryn Company, Cardiff and the Nos 3 and 4 pits ceased production in 1940, remaining open for maintenance purposes. After nationalisation in 1947, the National Coal Board drew up plans, approved in 1948, for a £5 million major modernisation scheme for working the coal reserves of the Cwmdare and Rhondda Fach valleys. A modern colliery was constructed on the site, using the existing Nos 3 and 4 shafts, and linking the colliery to the Bwllfa Colliery. A new mine was created to improve ventilation. Some output statistics from the new mine are shown below:

1954: 975 men, output 83,000 tons
1956: 901 men, output 166,000 tons
1958: 979 men, output 177,000 tons
1960: 1,502 men, output 312,000 tons.

After the strike in 1984-85, and given the political climate, total coal output was diverted via an underground roadway to Tower Colliery. No further coal was raised in the Rhondda Valley. When the pit finally closed on 21 December 1990 it was famous as the last working pit in the Rhondda Valleys.

In March 1996 the site was cleared, flattened and landscaped to provide a factory covering 160,000sqft and costing £7 million to build on the eight acre site. 250,000 tons of spoil was moved within the scheme boundary. The firm which took over the new factory site was Fenner Polymer, hoping to create 540 jobs. £25 million was invested in the project by Fenner, WDA, the Welsh Office, County and Borough Councils, and Mid-Glam. Tech. The factory uses the most up to date manufacturing technology to produce various rubber products for the automobile industry. The new plant was officially opened in December 1997, by the chief executive of the Welsh Development Agency, Mr. Brian Willet, who unveiled a plaque to mark the occasion. The chief executive of Fenner described the plant as a flagship of the company, and considered it as the most advanced facility of its kind in Europe.

Unfortunately the plant experienced early commissioning problems, currency fluctuations, and loss of orders through a general weakening in the automotive market, and was forced to make over 130 workers redundant in its first year. However, the Directors were confident that with an increase in production efficiency and an improvement in market forces they would attract further investment in the site, resulting in greater stability.

Mardy Colliery, c.1910.

Mardy Colliery in 1939.

A view of the 'Modern' colliery in the 1950s.

A panoramic view showing the new factory on the site of the colliery, and the village of Maerdy.

Above and below: The Memorial marking the closure of Mardy Colliery, and the end of 150 years of mining in the Rhondda Valleys. It is also a memorial to those who died while working at this colliery.

Mardy Colliery, Maerdy, 23 December 1885

The explosion took place without any warning at around 2.45p.m. The manager at the time, Mr W. Thomas called in Mr Thomas Wales, H.M. Inspector of Mines, who took charge of the rescue operation. No one realised that the death toll would reach eighty-one.

While records show that seventy-six miners were killed in the disaster, it appears that five of the severely injured who were rescued later died, bringing the total to eighty-one.
Those injured were:

Richard Davies
David Jones
? Lewis
John Henry Thomas
William Williams
Harry Isaac
John Jones
Gomer Rees
John Williams

The names of the dead are:

Joseph Baber, 17 years
John Bevan, 25 years
Arthur Boozey, 21 years
David Bowen, 18 years
John Collins, 40 years
Thomas Davies, 49 years
Thomas Davies, 13 years
Isaac Davies, 33 years
Evan Davies, 28 years
William Davies, 25 years
John Davies, 17 years
Evan Davies, 19 years
Lewis Davies, 32 years
John Edwards, 13 years
Edward Edwards, 52 years
Edward Edwards, 17 years
John Edwards, 45 years
David Evans, 16 years
Thomas Evans, 26 years
Thomas Evans, 26 years
John Evans, 25 years
John Evans, 55 years
Richard Evans, 24 years
Robert Griffiths, 34 years
William Griffiths, 16 years
William Harries, 29 years
John Heard, 22 years
Thomas Hughes, 33 years
Ephraim Hughes, 20 years
Philip Hutchins, 35 years
Evan James, 21 years
Thomas Jenkins, 25 years
John D. Jones, 51 years
David Jones, 59 years
William Jones, 16 years
David Jones, 25 years
David Jones, 55 years
Isaac Jones, 20 years
Joseph Jones, 40 years
William Jones, 13 years
John Jones, 42 years
David Jones, 27 years
David Lake, 35 years
Richard Lewis, 46 years
David Lewis, 40 years
John Lewis, 19 years
John Morgan, 17 years
Edmund Morgan, 27 years
Gomer Morgan, 21 years
James Parry, 29 years
Thomas Phillips, 25 years
Benjamin Phillips, 40 years
David Phillips, 50 years
Meshach Phillips, 32 years
John Powell, 23 years
Owen Powell, 29 years
Evan Pugh, 17 years
Henry Pullin, 23 years
Phillip Richards, 66 years
Evan Roberts, 28 years
David Rowlands, 27 years
Griffith Scourfield, 19 years
Joseph Spiller, 22 years
Michael Stokes, 17 years
James Sutton, 28 years
William Thomas, 19 years
David Thomas, 19 years
James Thomas, 24 years
Thomas Thomas, 24 years
Edward Thomas, 23 years
Owen Tudor, 32 years
Morgan Watkins, 14 years
Thomas Watkins, 17 years
Daniel Williams, 43 years
John Williams, 25 years
Levi Williams, 60 years.

The memorial to the eighty-one who died in the Mardy Colliery disaster of 1885, situated in the garden of Maerdy Church.